## ABOUT THE COVER



[Golden Jellyfish]

### JELLYFISH LAKE, PALAU

Jellyfish Lake is a saltwater lake located on Eil Malk island. This island is part of the Republic of Palau, which is an island nation in the Pacific Ocean. Jellyfish Lake is so named because it is home to many different species of jellyfish. One of the species is the golden jellyfish, which is shown in the picture.

Millions of golden jellyfish migrate across Jellyfish Lake each day. They are searching for food, which often is zooplankton found near the surface of the lake.

Jellyfish Lake is a snorkeling-only lake. Scuba diving is not allowed because bubbles from scuba tanks can be harmful if the bubbles collect beneath the bell of the jellyfish.

NATIONAL GEOGRAPHIC
SCIENCE

School Publishing

PROGRAM AUTHORS

Randy Bell, Ph.D.
Malcolm B. Butler, Ph.D.
Kathy Cabe Trundle, Ph.D.
Judith S. Lederman, Ph.D.
David W. Moore, Ph.D.

# Program Authors

**RANDY BELL, PH.D.**

Associate Professor of Science Education,
University of Virginia, Charlottesville, Virginia
SCIENCE

**MALCOLM B. BUTLER, PH.D.**

Associate Professor of Science Education,
University of South Florida, St. Petersburg, Florida
SCIENCE

**KATHY CABE TRUNDLE, PH.D.**

Associate Professor of Early Childhood Science Education, The School of Teaching and Learning,
The Ohio State University, Columbus, Ohio
SCIENCE

**JUDITH SWEENEY LEDERMAN, PH.D.**

Director of Teacher Education,
Associate Professor of Science Education,
Department of Mathematics and Science Education,
Illinois Institute of Technology, Chicago, Illinois
SCIENCE

**DAVID W. MOORE, PH.D.**

Professor of Education,
College of Teacher Education and Leadership,
Arizona State University, Tempe, Arizona
LITERACY

# Program Reviewers

**Amani Abuhabsah**
Teacher
Dawes Elementary
Chicago, IL

**Maria Aida Alanis, Ph.D.**
Elementary Science
Instructional Coordinator
Austin Independent School District
Austin, TX

**Jamillah Bakr**
Science Mentor Teacher
Cambridge Public Schools
Cambridge, MA

**Gwendolyn Battle-Lavert**
Assistant Professor of Education
Indiana Wesleyan University
Marion, IN

**Carmen Beadles**
Retired Science Instructional Coach
Dallas Independent School District
Dallas, TX

**Andrea Blake-Garrett, Ed.D.**
Science Educational Consultant
Newark, NJ

**Lori Bowen**
Science Specialist
Fayette County Schools
Lexington, KY

**Pamela Breitberg**
Lead Science Teacher
Zapata Academy
Chicago, IL

Program Reviewers continued on page iv.

---

## Acknowledgments

Grateful acknowledgment is given to the authors, artists, photographers, museums, publishers, and agents for permission to reprint copyrighted material. Every effort has been made to secure the appropriate permission. If any omissions have been made or if corrections are required, please contact the Publisher.

**Illustrator Credits**
All illustrations by Precision Graphics. All maps by Mapping Specialists.

**Photographic Credits**
**Front Cover** David Doubilet/National Geographic Image Collection.

Credits continue on page EM10.

Neither the Publisher nor the authors shall be liable for any damage that may be caused or sustained or result from conducting any of the activities in this publication without specifically following instructions, undertaking the activities without proper supervision, or failing to comply with the cautions contained herein.

**The National Geographic Society**
John M. Fahey, Jr.,
President & Chief Executive Officer

Gilbert M. Grosvenor,
Chairman of the Board



**National Geographic School Publishing**
Hampton-Brown
www.myNGconnect.com

Printed in the USA.
RR Donnelley
Jefferson City, MO

ISBN: 978-0-7362-7756-3

11 12 13 14 15 16 17 18 19 20

4 5 6 7 8 9 10

**Carol Brueggeman**
K–5 Science/Math Resource Teacher
District 11
Colorado Springs, CO

**Miranda Carpenter**
Teacher, MS Academy Leader
Imagine School
Bradenton, FL

**Samuel Carpenter**
Teacher
Coonley Elementary
Chicago, IL

**Diane E. Comstock**
Science Resource Teacher
Cheyenne Mountain School District
Colorado Springs, CO

**Kelly Culbert**
K–5 Science Lab Teacher
Princeton Elementary
Orange County, FL

**Karri Dawes**
K–5 Science Instructional Support Teacher
Garland Independent School District
Garland, TX

**Richard Day**
Science Curriculum Specialist
Union Public Schools
Tulsa, OK

**Michele DeMuro**
Teacher/Educational Consultant
Monroe, NY

**Richard Ellenburg**
Science Lab Teacher
Camelot Elementary
Orlando, FL

**Beth Faulkner**
Brevard Public Schools
Elementary Training Cadre,
Science Point of Contact, Teacher, NBCT
Apollo Elementary
Titusville, FL

**Kim Feltre**
Science Supervisor
Hillsborough School District
Newark, NJ

**Judy Fisher**
Elementary Curriculum Coordinator
Virginia Beach Schools
Virginia Beach, VA

**Anne Z. Fleming**
Teacher
Coonley Elementary
Chicago, IL

**Becky Gill, Ed.D.**
Principal/Elementary Science Coordinator
Hough Street Elementary
Barrington, IL

**Rebecca Gorinac**
Elementary Curriculum Director
Port Huron Area Schools
Port Huron, MI

**Anne Grall Reichel Ed. D.**
Educational Leadership/ Curriculum and Instruction Consultant
Barrington, IL

**Mary Haskins, Ph.D.**
Professor of Biology
Rockhurst University
Kansas City, MO

**Arlene Hayman**
Teacher
Paradise Public School District
Las Vegas, NV

**DeLene Hoffner**
Science Specialist, Science Methods Professor,
Regis University
Academy 20 School District
Colorado Springs, CO

**Cindy Holman**
District Science Resource Teacher
Jefferson County Public Schools
Louisville, KY

**Sarah E. Jesse**
Instructional Specialist for Hands-on Science
Rutherford County Schools
Murfreesboro, TN

**Dianne Johnson**
Science Curriculum Specialist
Buffalo City School District
Buffalo, NY

**Kathleen Jordan**
Teacher
Wolf Lake Elementary
Orlando, FL

**Renee Kumiega**
Teacher
Frontier Central School District
Hamburg, NY

**Edel Maeder**
K-12 Science Curriculum Coordinator
Greece Central School District
North Greece, NY

**Trish Meegan**
Lead Teacher
Coonley Elementary
Chicago, IL

**Donna Melpolder**
Science Resource Teacher
Chatham County Schools
Chatham, NC

**Melissa Mishovsky**
Science Lab Teacher
Palmetto Elementary
Orlando, FL

**Nancy Moore**
Educational Consultant
Port Stanley, Ontario, Canada

**Melissa Ray**
Teacher
Tyler Run Elementary
Powell, OH

**Shelley Reinacher**
Science Coach
Auburndale Central Elementary
Auburndale, FL

**Kevin J. Richard**
Science Education Consultant,
Office of School Improvement
Michigan Department of Education
Lansing, MI

**Cathe Ritz**
Teacher
Louis Agassiz Elementary
Cleveland, OH

**Rose Sedely**
Science Teacher
Eustis Heights Elementary
Eustis, FL

**Robert Sotak, Ed.D.**
Science Program Director,
Curriculum and Instruction
Everett Public Schools
Everett, WA

**Karen Steele**
Teacher
Salt Lake City School District
Salt Lake City, UT

**Deborah S. Teuscher**
Science Coach and Planetarium Director
Metropolitan School District of Pike Township
Indianapolis, IN

**Michelle Thrift**
Science Instructor
Durrance Elementary
Orlando, FL

**Cathy Trent**
Teacher
Ft. Myers Beach Elementary
Ft. Myers Beach, FL

**Jennifer Turner**
Teacher
PS 146
New York, NY

**Flavia Valente**
Teacher
Oak Hammock Elementary
Port St. Lucie, FL

**Deborah Vannatter**
District Coach, Science Specialist
Evansville Vanderburgh School Corporation
Evansville, IN

**Katherine White**
Science Coordinator
Milton Hershey School
Hershey, PA

**Sandy Yellenberg**
Science Coordinator
Santa Clara County Office of Education
Santa Clara, CA

**Hillary Zeune de Soto**
Science Strategist
Lunt Elementary
Las Vegas, NV

# LIFE SCIENCE

## CONTENTS

## How Do Living Things Depend on Their Environment? ........ 85

CHAPTER 4

## How Do Adaptations Help Living Things Survive? ........ 133



CHAPTER 5

## How Do Living Things Interact with Their Environment? ........ 181

# TECHTREK
myNGconnect.com

Student eEdition

Vocabulary Games

Digital Library

Enrichment Activities

CHAPTER 6

LIFE
SCIENCE

# What Is Life Science?

Life science is the study of all the living things around you and how they interact with one other and with the environment. This type of science investigates how living things are similar to and different from one another, how they live and reproduce, and how they function in the environment. Life science includes the study of humans, as well as all the other kinds of living things on Earth. People who study living things and the environment are called life scientists.

You will learn about these aspects of life science in this unit:

## HOW DO PLANTS GROW AND REPRODUCE?

**Plants go through life cycles that result in more plants of the same kind. Plants inherit characteristics from the parent plants, although some characteristics may be shaped by the environment.**

## HOW DO ANIMALS GROW AND CHANGE?

**Animals go through life cycles in which many changes occur. Some of these changes are shaped by characteristics inherited from the parents, while others are a result of the environment in which animals are growing and reproducing.**

## HOW DO LIVING THINGS DEPEND ON THEIR ENVIRONMENT?

**The environment includes both living and nonliving parts. The nonliving parts of the environment determine what kinds of living things are there. But the living things in all environments interact with one another, moving energy from the sun through all living things. Life scientists study how living things depend on the environment.**

## HOW DO ADAPTATIONS HELP LIVING THINGS SURVIVE?

**Plants and animals have adaptations that help them move, get food, and even reproduce. Adaptations can be body parts or plant parts. Some adaptations are behaviors, such as communication.**

## HOW DO LIVING THINGS INTERACT WITH THEIR ENVIRONMENT?

**Seasonal changes in the environment cause changes in plants and animals. Plants and animals cause changes in the environment too. The actions of humans and all animals have effects—large and small—on their surroundings.**

## HOW DO THE PARTS OF AN ORGANISM WORK TOGETHER?

**Humans have many body systems. Each body system does a job in the body. Many different organs make up the different systems. Life scientists study how body systems work together.**

# MEET A SCIENTIST

## Luke Dollar: Conservation Scientist

Luke Dollar is a conservation scientist and National Geographic Emerging Explorer. Luke is devoted to ensuring that every minute and dollar is spent on conservation efforts that really count. "I take the money I raise or borrow straight to the ground level and get more bang for my buck," reports Luke.

Luke's recent focus has been on researching and studying the elusive predator found only on the island of Madagascar—the fosa. As one of the most important species on the island, the fosa plays a crucial role in maintaining the equilibrium of Madagascar's entire food chain.

Luke's research has two parts: tracking the hard-to-find fosa and monitoring forests. To carry out his research, Luke and his team observe and count fosa.

## After reading Chapter 1, you will be able to:

- Identify the major structures of a seed plant and relate them to their functions. **HOW PLANTS GROW**
- Classify plants according to their characteristics. **HOW PLANTS GROW, LIFE CYCLES OF SEED PLANTS**
- Describe the main processes involved in the reproduction of flowering plants. **FLOWERS, FRUITS AND SEEDS**
- Compare and contrast the life cycles of flowering plants and conifers. **LIFE CYCLES OF SEED PLANTS**
- Identify individual differences in organisms of the same kind. **CHARACTERISTICS OF PLANTS**
- Explain that although most characteristics of plants are inherited, some characteristics can be affected by factors in the environment. **CHARACTERISTICS OF PLANTS**
- Science in a Snap! Describe the main processes involved in the reproduction of flowering plants. **FRUITS AND SEEDS**

CHAPTER

# 1

# HOW DO PLANTS AND

**Have you ever** looked closely at a dandelion? Every yellow “flower” is made up of many smaller flowers. Each of these flowers forms a seed. A silky parachute is attached to each seed. When the wind blows, the parachutes carry the seeds away. How do these parachutes help the dandelion reproduce?

# GROW REPRODUCE?

Each dandelion seed holds a new plant and food to help it start to grow.

# SCIENCE VOCABULARY

## photosynthesis
(FŌ-tō-SIN-thuh-sis)

In **photosynthesis**, plants use the energy of sunlight to make food. (p. 11)

Green leaves carry out photosynthesis.

## conifer (CON-uh-fur)

A **conifer** is a seed plant that reproduces with cones. (p. 12)

A pine tree is a conifer.

## pollination
(POL-uh-NĀ-shun)

**Pollination** is the movement of pollen from a stamen to a pistil or from a male cone to a female cone. (p. 16)

When a butterfly carries pollen from flower to flower, pollination occurs.

**conifer**
(CON-uh-fur)

**fertilization**
(FUR-til-i-ZĀ-shun)

**inherited**
(in-HAIR-it-ed)

**photosynthesis**
(FŌ-tō-SIN-thuh-sis)

**pollination**
(POL-uh-NĀ-shun)

**seed dispersal**
(SĒD di-SPUR-suhl)

Vocabulary Games

## fertilization
(FUR-til-i-ZĀ-shun)

In **fertilization**, an egg and a sperm cell join. (p. 17)

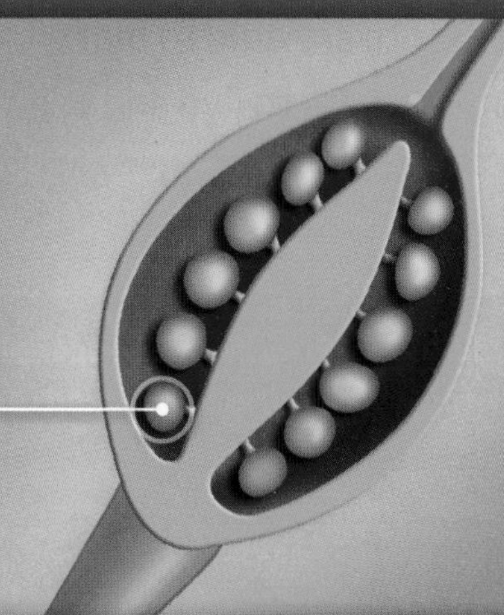

Fertilization takes place in the ovule of a flower.

## seed dispersal
(SĒD di-SPUR-suhl)

In **seed dispersal**, the seeds of a plant are carried to a new place. (p. 20)

Wind helps in the seed dispersal of dandelions.

## inherited (in-HAIR-it-ed)

An **inherited** characteristic is passed down from parents to offspring. (p. 32)

The color of each tulip was inherited from its parents.

# How Plants Grow

Have you ever looked closely at a plant? If so, you know it is made up of leaves, stems, and roots. It may also have flowers and fruits. Each of these parts carries out a different job. But they all work together to help the plant grow and reproduce.

Like all living things, plants need food for energy. But unlike animals, most plants are able to make their own food. Plants use food and other nutrients to live and grow larger.

The different parts of this pepper plant work together to help it grow and reproduce.

Plants make food by **photosynthesis**. Photosynthesis uses the energy of sunlight to change water and carbon dioxide into food. The water comes from the soil. The carbon dioxide comes from the air.

To carry out photosynthesis, leaves use a green substance called chlorophyll. Chlorophyll is able to capture the energy of sunlight. Leaves use this energy to combine water and carbon dioxide, making food. This process gives off oxygen. The oxygen goes into the air.

**Sunlight + Water + Carbon Dioxide → Food + Oxygen**

**Leaves make food for a plant by photosynthesis.**

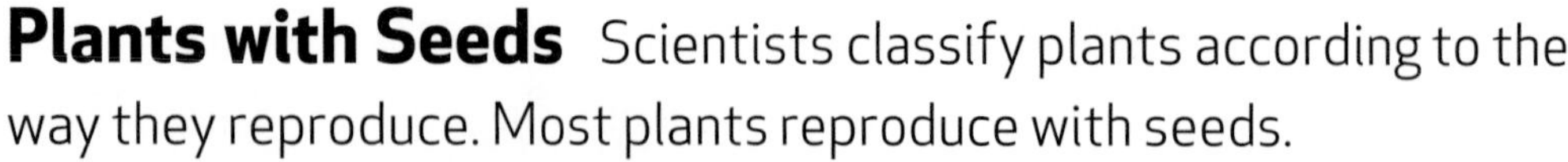

**Plants with Seeds** Scientists classify plants according to the way they reproduce. Most plants reproduce with seeds.

The largest group of seed plants are the flowering plants. Roses and daisies are flowering plants. You may be surprised to learn that grasses have flowers, too. Trees that have broad leaves, such as oaks, also have flowers. The seeds of all flowering plants grow in fruits.

A second group of seed plants do not have flowers. These plants form their seeds in cones. Plants that have cones are called **conifers**. Pine trees are conifers. Conifers usually have leaves shaped like needles or scales.

**Plants without Seeds** Ferns and mosses do not form seeds. These plants reproduce with spores. Spores are much smaller than seeds. Spores do not carry food for the young plant.

You may have seen ferns growing on the forest floor. Like seed plants, ferns have roots, stems, and leaves.

Mosses are usually much smaller than ferns. They do not have roots or stems to carry water and nutrients to the rest of the plant. Instead, mosses grow low to the ground.

This tree fern grows in the damp woodlands of Tasmania.

## Before You Move On

1. What two groups of plants reproduce with seeds?
2. How do leaves make food for the plant?
3. **Infer** Spruce trees have leaves that are shaped like needles. Do you think spruce trees reproduce with flowers or cones? Explain your reasoning.

# Flowers

**Parts of a Flower** Flowers such as the lily below are beautiful. But plants do not have flowers to look pretty. Plants have flowers to make seeds and fruits. To understand how, you need to learn the parts of a flower.

The first thing you notice about most flowers is their petals. The color and shape of the petals attract insects and other animals to the flower.

The bright color and sweet smell of this lily attract insects.

Now look at the center of the flower. Can you see the pistil and stamens? The pistil is the female part of the flower. The base of the pistil is called the ovary. Inside the ovary are many ovules that contain eggs.

The stamens are the male parts of the flower. Notice how the tips of the stamens are orange. They are orange because they make an orange powder called pollen.

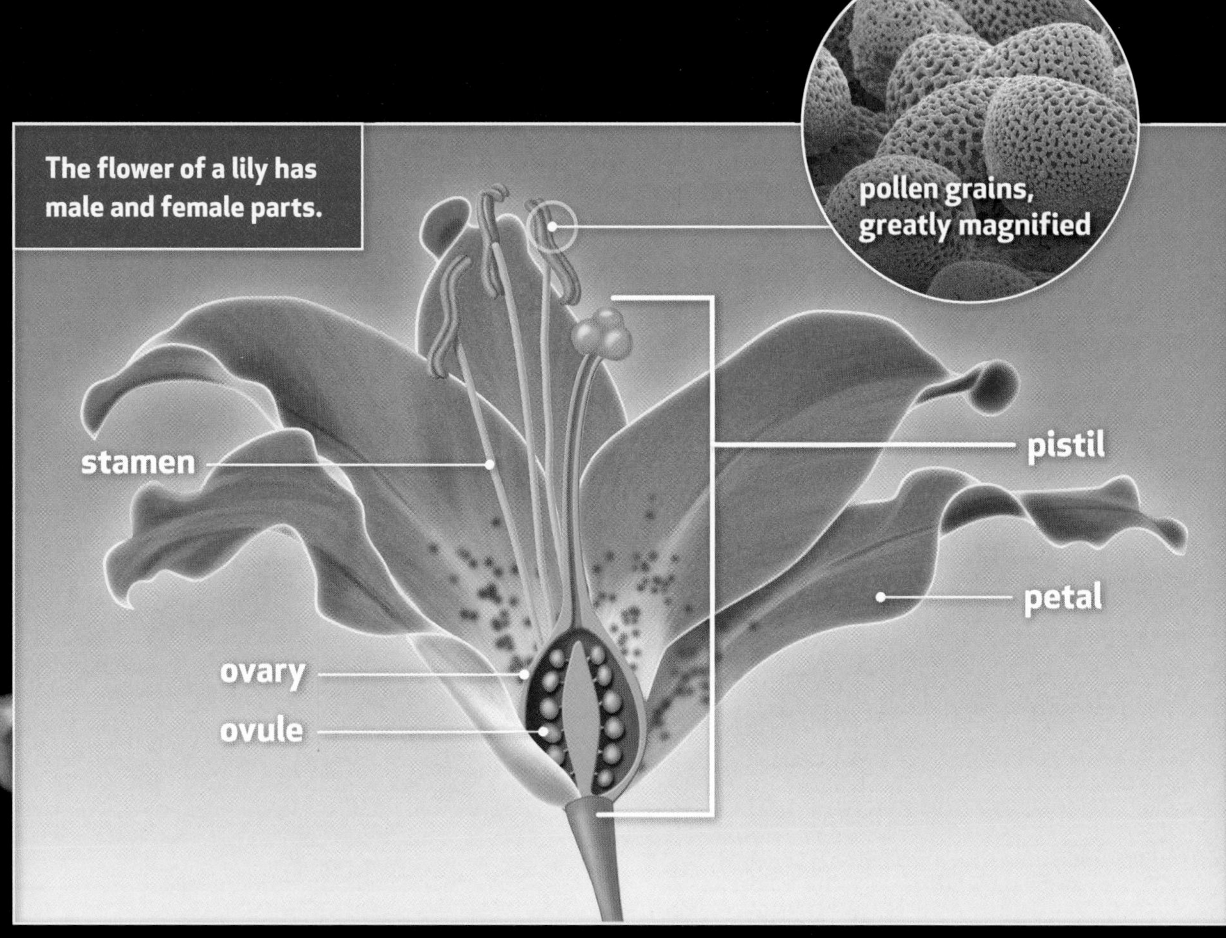


The flower of a lily has male and female parts.

**Pollination** Deep inside many flowers is a sweet nectar. Bees and many other insects visit flowers to drink this nectar. As an insect looks for nectar, its body brushes by the stamens. Grains of pollen stick to its body. When the insect visits another flower, some of this pollen is left on the pistil of the second flower. The movement of pollen from a stamen to a pistil is called **pollination**.

Insects are not the only animals that pollinate flowers. Some flowers are pollinated by birds, or even bats. Most grasses and trees are pollinated by the wind. Wind blows the pollen from one flower to the pistil of another flower.

As this butterfly searches for nectar, it carries pollen from one orchid to another.

**Fertilization** In pollination, a grain of pollen lands on the sticky tip of a pistil. Soon a thin tube grows from the pollen grain. The pollen tube grows down the pistil to the ovary. Inside the ovary are ovules. Each ovule holds many eggs. When the pollen tube reaches an ovule, a sperm cell joins with an egg. The joining of an egg and sperm cell is called **fertilization**.

After fertilization, the flower begins to change. Seeds and fruits begin to form. The petals dry up and fall off.

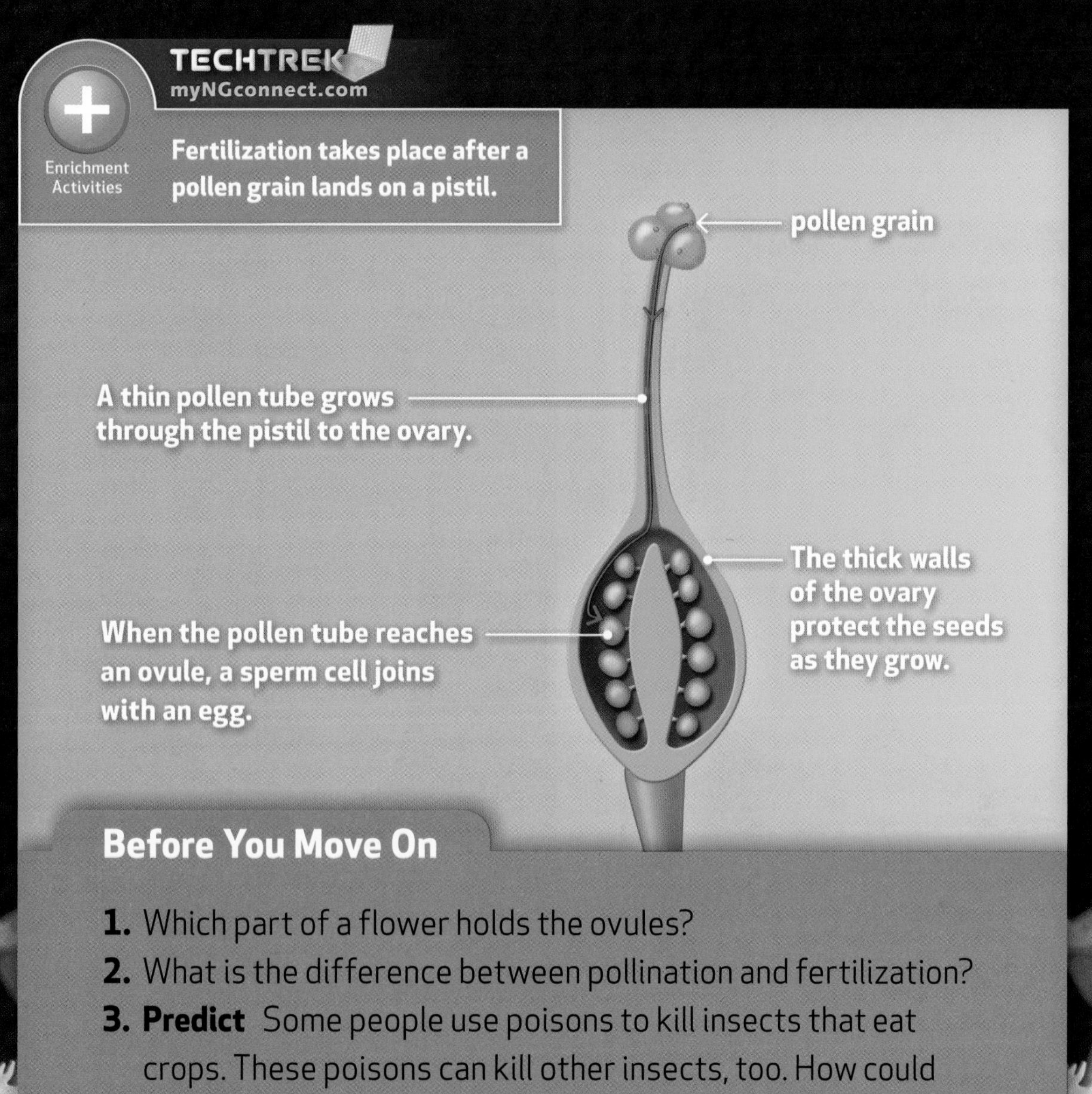


Fertilization takes place after a pollen grain lands on a pistil.

## Before You Move On

1. Which part of a flower holds the ovules?
2. What is the difference between pollination and fertilization?
3. **Predict** Some people use poisons to kill insects that eat crops. These poisons can kill other insects, too. How could these poisons affect the reproduction of flowering plants?

# Fruits and Seeds

**Seeds** After fertilization, seeds and fruits begin to grow. Seeds develop from ovules inside the ovaries.

Each seed holds everything that a new plant needs to grow. If you cut a seed open, you will find three main parts. These are the embryo, the seed coat, and stored food. The embryo will grow into a new plant. The seed coat protects the embryo and keeps it from drying out. The stored food helps the embryo to grow.

Look at the pictures of a corn seed and a bean seed. Both of these seeds have the three main parts. But the bean seed can split into two parts. The corn seed does not split into two parts. What other differences do you see?

A bean seed has two seed leaves. These thick leaves store food.

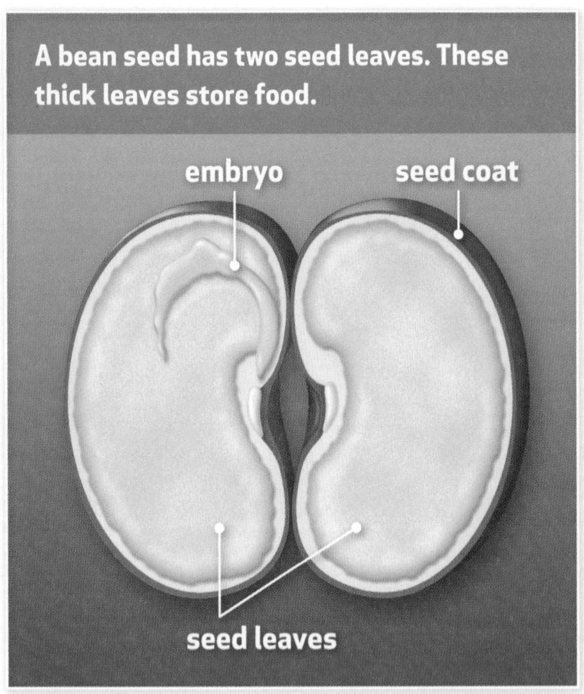


Corn seeds have only one seed leaf. Food is stored in another part of the seed.

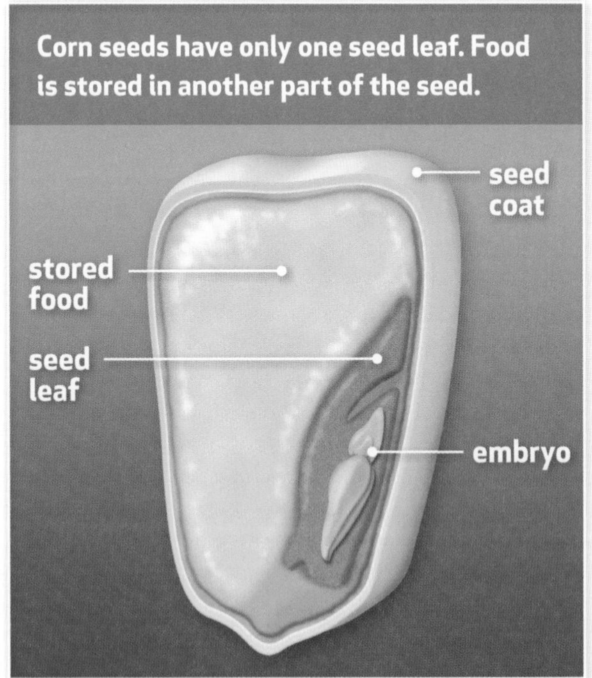

**Fruits** Fruits grow from the walls of the ovaries that surround the seeds. Fruits protect and help to spread the seeds.

There are many different kinds of fruits. Some fruits, such as cherries, hold one seed. Other fruits, such as apples and pears, hold several seeds. Pumpkins have many seeds.

Apples and peaches are sweet and fleshy. Other fruits are dry. The shell of a walnut is a dry fruit. The nut that you eat is the seed. The pods that surround beans and peas are also fruits. The beans and peas are seeds.

Fruits come in many shapes and sizes. Even tomatoes are fruits, although you may call them vegetables!

**Seed Dispersal** Seeds cannot move from place to place by themselves. For seeds to reach new places, something must carry them. In **seed dispersal**, the seeds of a plant are carried to a new place. Fruits help in seed dispersal.

Many seeds are spread by animals. When birds eat berries and other fleshy fruits, they also eat the hard seeds inside. Later the seeds drop to the ground. Squirrels gather nuts and bury them in the ground. If a squirrel does not find all the nuts, some may grow into new trees.

Some fruits have hooks that stick in an animal's fur. Later the seeds drop from the animal.

**This cedar waxwing is helping to spread the seeds in the pyracantha berries.**

Other seeds are carried by the wind. Dandelions have silky fruits that float on the wind. The seeds of some plants have tiny wings. These wings help the seeds spin through the air.

Some seeds are spread by water. Coconuts are the fruits of palm trees. Because coconuts float, ocean water can carry them far away. When coconuts wash onshore, they grow into new trees.

**The fruits of coconuts are spread by water.**

**Dandelion seeds are spread by the wind.**

## Science in a Snap! Observing Fruits

Observe a strawberry that has been cut in half. Draw what you see. Label the seeds.

Use a hand lens to carefully observe a maple fruit. Carefully pull the thick part of the fruit open. Draw and label what you see inside.

**Compare the fruits and seeds. Which seeds are dispersed by animals? Which are dispersed by wind? Explain your reasoning.**

**Germination and Growth** Have you ever planted a garden? When the weather is warm enough, you put seeds in the soil. Then you water them. Seeds will not grow unless they have enough water. But when conditions are right, a seed germinates, or starts to grow.

First, the seed takes in water and begins to swell. Next, the seed coat breaks open. The root starts to grow down. Then the stem and young leaves grow upward.

TECHTREK
myNGconnect.com
Digital Library

## GERMINATION OF A **BEAN SEED**

**A bean seed has two thick seed leaves. After the seedling pokes through the soil, other leaves begin to make food.**

**When a bean seed germinates, the seed leaves protect the stem and new leaves as they push up through the soil.**

A young plant is called a seedling. The seedling uses the food that was stored in the seed to start growing. When the new leaves form, they use sunlight to make food for the plant.

The seedling continues to increase in size. The stem grows taller and more leaves grow. In the soil, the roots grow deeper and wider. Over time, the seedling grows into a mature plant that is able to reproduce.

## GERMINATION OF A **CORN SEED**

**The seed leaf of a corn plant stays underground. A tough covering protects the new leaves as they grow through the soil.**

## Before You Move On

1. What part of a flower becomes a seed?
2. How does a fruit help a plant reproduce?
3. **Analyze** Why is seed dispersal important to a plant? What would happen if its seeds were not dispersed?

# WE NEED HONEYBEES!

Have you ever seen honeybees buzzing around an orchard? Honeybees visit flowers to get nectar. What do the bees do with the nectar they gather? They take it back to their hives to make honey.

As honeybees move from flower to flower, they carry pollen. This pollinates the flowers. Honeybees help pollinate many crops, such as almonds, apples, and strawberries. Without honeybees, these plants could not make much fruit.

**Bees visit flowers to get nectar and pollen.**

Beekeepers raise honeybees to help pollinate crops and to harvest honey. But lately many bees have died. Beekeepers are having trouble raising enough bees. What is causing the problem?

One answer is the spread of mites. Mites are tiny animals that live on the bodies of bees. Mites make it hard for bees to fly. Mites can also kill bees. Scientists are trying to find ways to protect bees from mites. Our food depends on bees!

When almonds blossom, beekeepers bring hives of bees to help pollinate the trees.

These mites live on the bodies of bees. Too many mites can kill a bee.

An almond grove in blossom.

# Life Cycles of Seed Plants

**Life Cycles of Flowering Plants** Oak trees can live for hundreds of years. Some wildflowers live for only a few months. But all flowering plants go through similar stages of life.

Follow the diagram as you read about the life cycle of an orange tree. An orange tree begins life as a seed. Then the seed germinates and the seedling begins to grow. Over time, the seedling grows into a mature tree that can flower. Bees and other insects pollinate the orange blossoms. After fertilization, seeds and fruit form. When these seeds germinate, the cycle begins again.

**Orange trees blossom in the spring. Their fruits are harvested in the fall.**

Orange trees can live for many years. Each year they flower and make new seeds. But after many years, an orange tree will die. Other kinds of flowering plants, such as corn and bean plants, die after they make seeds one time.

## LIFE CYCLE OF AN **ORANGE TREE**

**GERMINATION** When conditions are good, a seed will start to grow.

**GROWTH** The seedling grows into a young tree.

**MATURITY** When it is large enough, the tree will begin flowering. Bees carry pollen from one flower to another.

**SEED FORMATION** When a flower is fertilized, a fruit begins to grow. Each orange holds a number of seeds.

**Life Cycles of Conifers** The picture below shows different stages in the life cycle of a longleaf pine. How many different stages can you see in the picture?

Longleaf pines are conifers. Like flowering plants, conifers begin life as seeds. The seedlings of longleaf pines grow slowly. When the trees are mature, they reproduce by forming cones.

Pine trees have two kinds of cones—male cones and female cones. Male cones make pollen. Wind blows the pollen to the female cones. When pollen reaches a female cone, fertilization takes place. Seeds begin to grow inside the scales of the cone.

**Longleaf pines grow in open woodlands called pine savannas.**

The female cone protects the seeds as they grow. When the seeds are fully grown, the cone opens. Some seeds fall to the ground. Others are carried away by the wind. If a seed lands in a good place, it will germinate.

## LIFE CYCLE OF A **LONGLEAF PINE**

**GERMINATION** If a seed lands in a good place, it will start to grow.

**GROWTH** The seedling grows into a young tree.

**MATURITY** When it is large enough, the tree makes cones. Male cones produce pollen that is spread by the wind.

**SEED FORMATION** When pollen reaches the female cones, the ovules are fertilized. Seeds form inside the female cones.

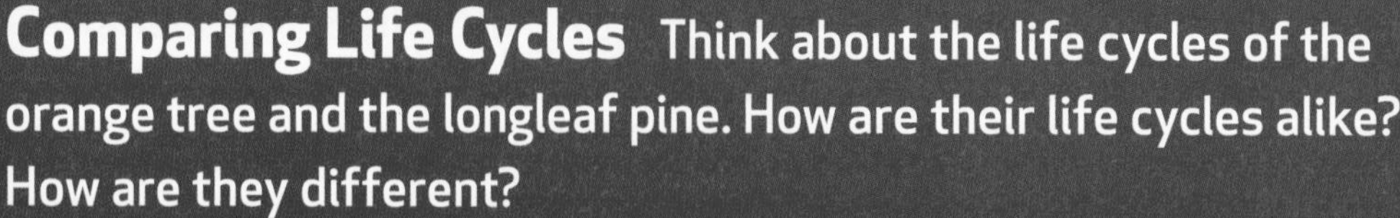

**Comparing Life Cycles** Think about the life cycles of the orange tree and the longleaf pine. How are their life cycles alike? How are they different?

The seeds of an orange tree are protected by the tough skin of its fruit.

The seeds of a longleaf pine are protected by the female cones.

Both kinds of plants reproduce with seeds. The seeds of conifers grow in cones. Conifer seeds are not protected by fruits. The seeds of flowering plants form in flowers. Their seeds are protected by fruits.

Conifers are pollinated by the wind. Some flowering plants are also wind pollinated. But most flowering plants are pollinated by insects or other animals.

All conifers are trees or shrubs that live for many years and reproduce many times. Some kinds of flowering plants also live for many years. But many kinds of flowering plants have much shorter life cycles. Some live for only a season. After they make seeds, they die.

## COMPARING **SEED PLANTS**

| | **REPRODUCTIVE STRUCTURES** | **POLLINATION** | **LOCATION OF SEEDS** |
|---|---|---|---|
| **FLOWERING PLANTS** | Flowers | Most are pollinated by animals.<br>Some are pollinated by wind. | In fruits |
| **CONIFERS** | Cones | All are pollinated by wind. | In cones |

### Before You Move On

1. What are the four main stages in the life cycle of a flowering plant?
2. Compare pollination in conifers with pollination in flowering plants.
3. **Apply** Which do you think would grow more quickly, a garden of flowering plants or a garden of conifers? Explain why.

# Characteristics of Plants

As you know, young plants are similar to their parents. The leaves of an orange tree are the same shape as those of its parents. The shape of an orange leaf is an **inherited** characteristic. Characteristics that are inherited have been passed from parents to their offspring.

Look at the tulips in the picture below. Why are the tulips different colors? The color of each flower has been inherited from different parent flowers.

The color of each of these tulips was inherited from the plant's parents.

Most characteristics of plants are inherited. But things around a plant can affect the way it grows. Characteristics that have been changed by the environment cannot be passed to offspring.

The environment can change plants in many ways. If a plant does not get enough sunlight, its leaves may turn yellow or grow very large. Nutrients in the soil also affect how plants grow. The plants below were grown in different kinds of soil. How do these plants differ?

**These two petunia plants are the same age. The one on the left was grown in soil that had few nutrients.**

## Before You Move On

1. What does the word *inherited* mean?
2. Use an example to explain how the environment can affect the way a plant grows.
3. **Draw Conclusions** You have a package of flower seeds. You plant some seeds in the shade and other seeds in sunlight. Later you observe that plants in the shade have longer stems than plants growing in sunlight. Is this difference inherited? Why?

The two main groups of plants are those that reproduce with seeds and those that do not have seeds. Flowering plants and conifers are seed plants. Flowering plants reproduce with seeds that grow in fruits. Conifers reproduce by making seeds in cones. Most of a plant's characteristics are inherited. But some characteristics are changed by its environment.

**Big Idea** The life cycle of a plant includes the germination of a seed, growth, maturity, reproduction, and finally death.

**Seeds Germinate**

**Plants Grow and Reproduce**

**Plants Die**

## Vocabulary Review

**Match** the following terms with the correct definition.

**A. conifer**
**B. photosynthesis**
**C. pollination**
**D. fertilization**
**E. inherited**
**F. seed dispersal**

1. The joining of an egg and a sperm cell
2. The carrying of seeds to a new place
3. A seed plant that reproduces with cones
4. The movement of pollen from a stamen to a pistil
5. A characteristic that is passed from parents to offspring
6. The way plants use the energy of sunlight to make food

## Big Idea Review

1. **Identify** What part of a plant carries out photosynthesis?
2. **Describe** List the three parts of a seed. Describe how each part helps the new plant live and grow.
3. **Explain** What are the two kinds of pine cones? How is each involved in the reproduction of a pine tree?
4. **Sequence** Put these steps in order, beginning with flowering: flowering, germination, fertilization, seed dispersal, pollination.
5. **Generalize** Are animals important in the reproduction of flowering plants? Give reasons for your answer.
6. **Infer** The fruits of the burdock plant have many little hooks. How do you think the seeds of the burdock are dispersed?

## Write About Life Cycles

**Explain** Describe the main stages in the life cycle of a squash plant growing in a garden. Use the words *germinate* and *reproduce* in your description.

CHAPTER 1

# LIFE SCIENCE EXPERT: PLANT BIOLOGIST

## Mark Olson: Plant Biologist

Mark Olson is a plant biologist who teaches at Mexico's National University. He studies how plants develop different shapes and sizes in different parts of the world.

### As a plant biologist, what do you study?

Almost any place on Earth is covered with many different living things. In the dry tropical shrubland outside my office, there are very large shrubs and tiny herbs. Vines cling to the branches of the other plants. This mixture in shapes, sizes, and ways of living is what makes our Earth so interesting. The question I study is: Where did all this diversity come from?

### How do you study plants in the field?

I spend a little more than half of my time working in the field. We may spend the day measuring tiny cacti on a desert plain, collecting plants on the cliffs above a tropical beach, or climbing a mountain to find a rare plant. We might use ropes to climb rain forest trees. Sometimes we fly above the rain forest in a glider to learn how the trees collect light from the sun.

Mark Olson is wearing a helmet to protect him as he flies in the glider.

Mark studies the shape of plants such as this cactus.

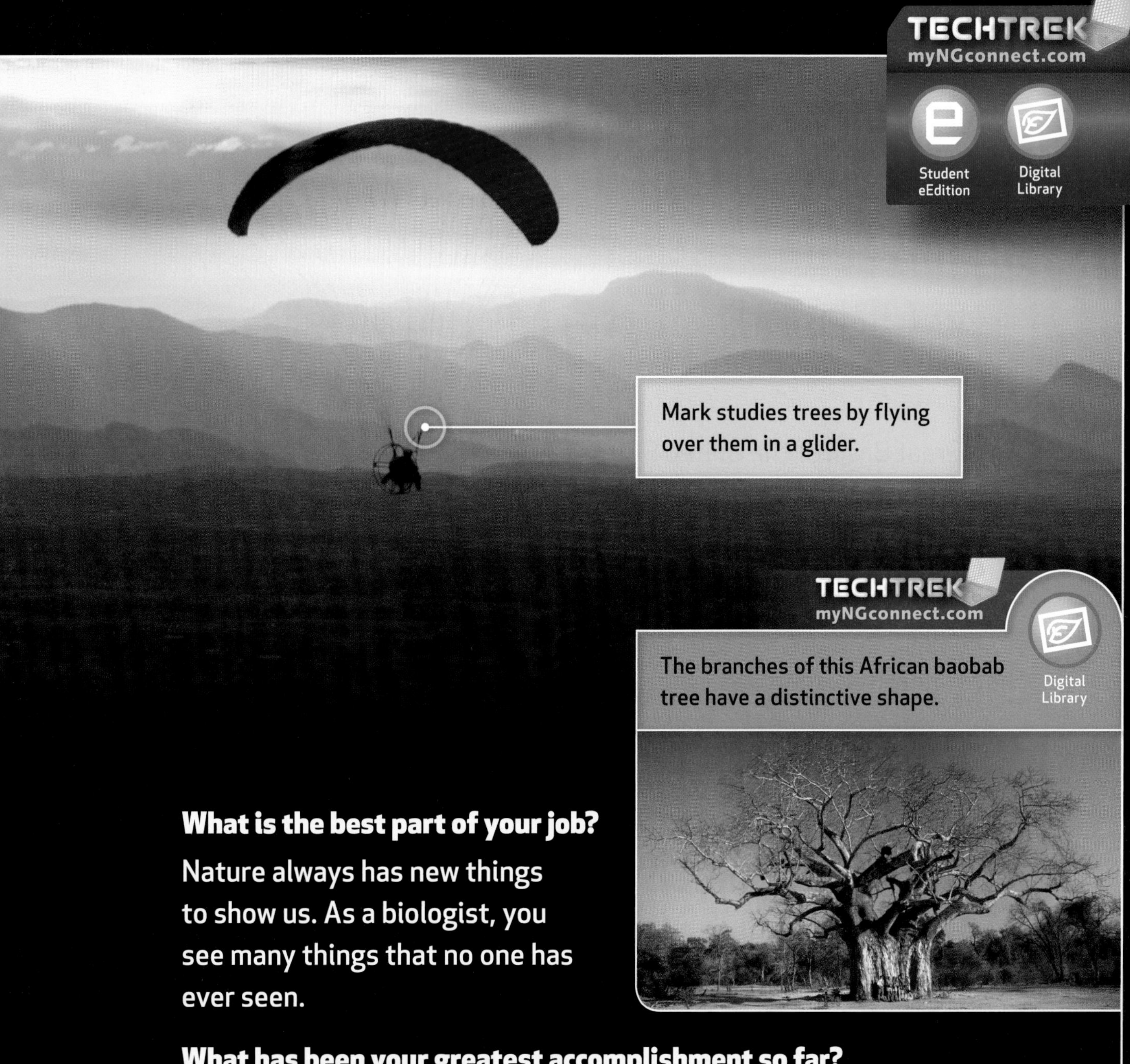


Mark studies trees by flying over them in a glider.



The branches of this African baobab tree have a distinctive shape.

**What is the best part of your job?**

Nature always has new things to show us. As a biologist, you see many things that no one has ever seen.

**What has been your greatest accomplishment so far?**

Trees around the world have a pattern in their shapes. Trees with small leaves have small branches that branch frequently. Trees with large leaves have thick twigs that branch sparingly. My team found that this pattern is caused by leaves, stems, and roots working together.

**What skills would someone need to have if they wanted to be successful at what you do?**

The most important skill is simply to work hard. Scientists work hard to collect the information they need to draw conclusions.

NATIONAL GEOGRAPHIC

BECOME AN EXPERT

# Bristlecone Pines: The Old Ones of the Mountains

## Living in a Harsh Land

The Great Basin of the United States is a desert. Summers are hot and dry. Winters are cold and windy. Here, rugged mountains rise from flat valleys. Near the top of these mountains, you might find a windy ridge with trees that are twisted and bent. These are bristlecone pines.

Most plants cannot live in this harsh land. But bristlecone pines have **inherited** characteristics that allow them to grow here. Their roots can grow in the dry, rocky soil. Their short, stiff needles can live through the long, cold winter. In the summer, their needles make food by **photosynthesis**.

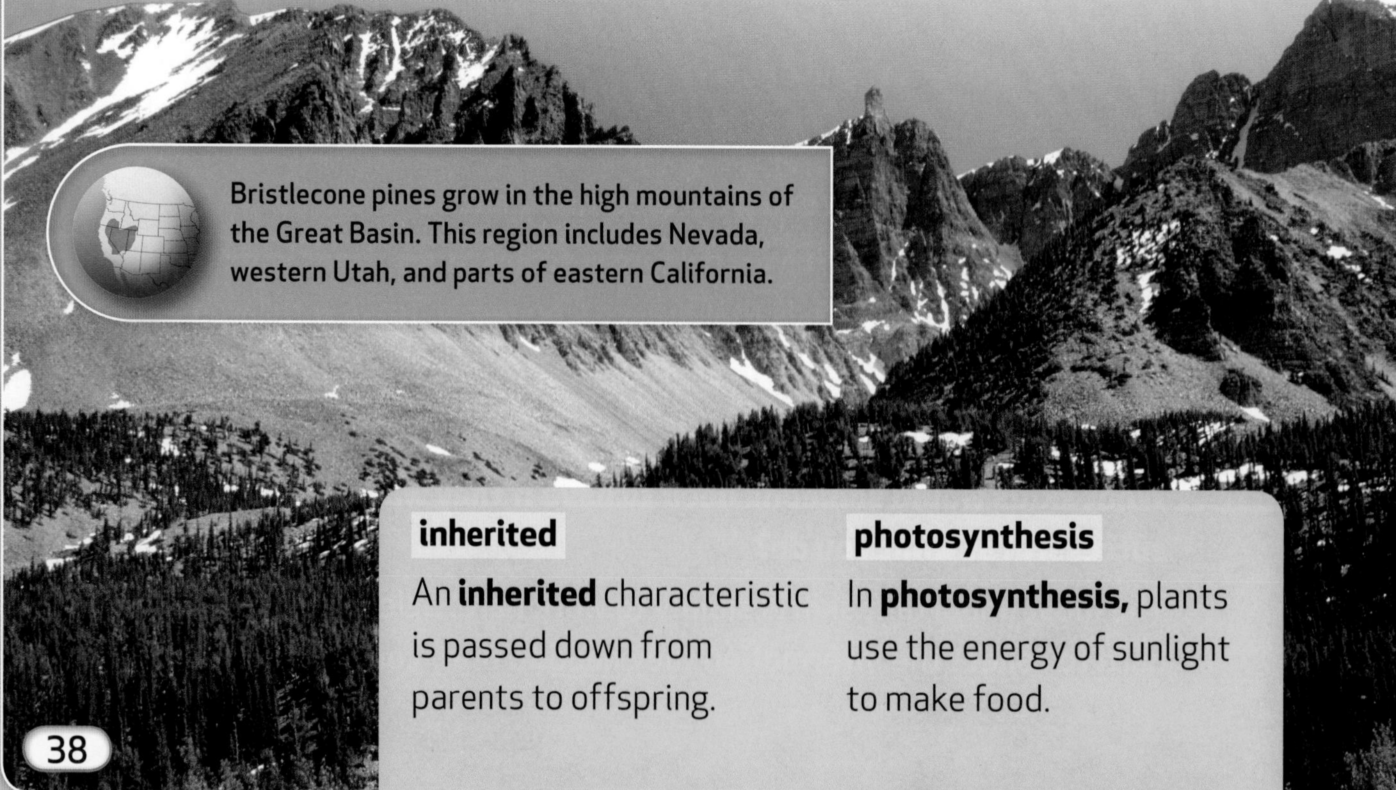

Bristlecone pines grow in the high mountains of the Great Basin. This region includes Nevada, western Utah, and parts of eastern California.

**inherited**

An **inherited** characteristic is passed down from parents to offspring.

**photosynthesis**

In **photosynthesis,** plants use the energy of sunlight to make food.

Bristlecone pines also grow lower on the mountains, where it is warmer and less windy. There the tree trunks grow straight up.

Near the top of the mountains, the winds are stronger. These cold, dry winds change the shape of the trees. Instead of growing upward, the bristlecones bend sideways, away from the wind. Here the trees grow very slowly. Even though they are not very tall, they are not young. Some are thousands of years old!

TECHTREK
myNGconnect.com
Digital Library

**Strong winds have twisted the trunk and branches of this bristlecone pine.**

## Bristlecones Old and New

Even the oldest pine began life as a seed. That seed grew in a cone.

Bristlecones have two kinds of cones—male and female. Male cones make large amounts of pollen. Wind blows the pollen away. If a pollen grain lands on a female cone, **pollination** takes place.

Next, a tube grows from the pollen grain. When it reaches an ovule, a sperm cell joins with an egg. After **fertilization**, a seed begins to grow. The female cone protects the seeds.

The male cones are red. Wind moves pollen from the male cones to the female cones.

The young female cones are purple or blue. Notice the tiny bristles on the female cone. These bristles give the trees their name!

**pollination**

**Pollination** is the movement of pollen from a stamen to a pistil or from a male cone to a female cone.

**fertilization**

In **fertilization,** an egg and a sperm cell join.

When the seeds are ripe, the cones open. The seeds have small wings, which the wind carries away.

Birds called Clark's nutcrackers are important in the **seed dispersal** of bristlecone pines. These birds gather the seeds and bury them in the ground. Later the nutcrackers return and eat the seeds. But they do not find all the seeds that they buried.

If a seed is not eaten, a seedling may grow. The young bristlecone grows very slowly. After twenty to fifty years, it becomes a mature tree. The tree will make cones for as long as it lives.

The seedling grows slowly in the dry limestone soil.

A bristlecone pine germinating.

This Clark's nutcracker is collecting seeds from pine cones.

**seed dispersal**

In **seed dispersal,** the seeds of a plant are carried to a new place.

## How Long Do They Live?

The life cycle of a bristlecone starts with a seed. But once it is mature, a tree may live for thousands of years. Some bristlecone pines have lived for nearly 5,000 years. These ancient trees have only a tiny bit of living wood. Finally, even the oldest bristlecones die.

How does the life cycle of a bristlecone pine compare with that of other trees? Look at the chart below to find the answer. Are you surprised to see that these **conifers** live longer than trees that have flowers?

# WHICH TREES LIVE THE LONGEST?

**The chart compares the life spans of some American trees.**

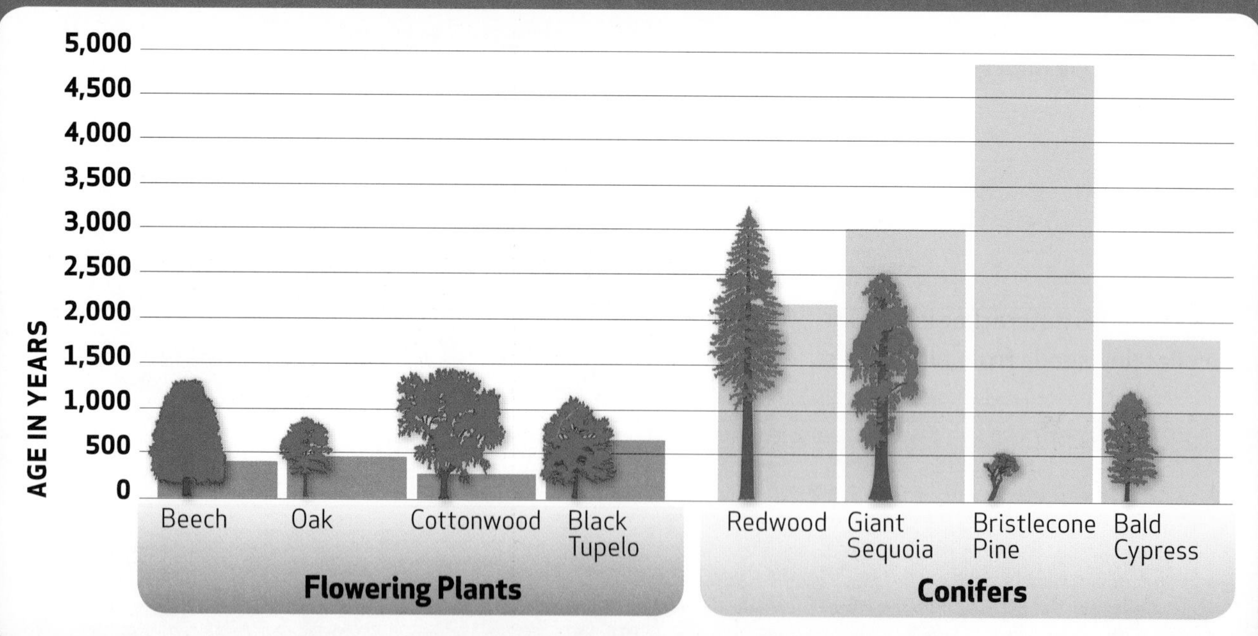


**conifer**

A **conifer** is a seed plant that reproduces with cones.

How do scientists learn how old a tree is? They look at its growth rings. Each year, the trunk of a tree grows a tiny bit wider. This growth forms a ring. Scientists count these rings to find how many years the tree has lived. If a tree is cut down, it is easy to see the rings. But scientists can also take a tiny core from a tree trunk and count the rings.

Scientists counted the rings on one bristlecone pine that had lived for 4,950 years. That tree was already 300 years old when the Pyramids of Egypt were being built!

**This scientist is using a drill to take a core from the trunk of a bristlecone.**

**The growth rings of this bristlecone tell how many years it lived. Because the tree grew slowly, its rings are close together.**

NATIONAL GEOGRAPHIC

# CHAPTER 1 SHARE AND COMPARE

**Turn and Talk** How does the environment affect the way that a bristlecone pine tree grows? Work with a partner to form a complete answer to this question.

**Read** Select two pages in this section. Practice reading the pages. Then read them aloud to a partner. Talk about why the pages are interesting.

my SCIENCE notebook **Write** Write a conclusion that tells the important ideas you learned about the life cycle of a bristlecone pine. State what you think is the Big Idea of this section. Share what you wrote with a classmate. Compare your conclusions. Did your classmate include all the stages of the life cycle?

my SCIENCE notebook **Draw** Draw a picture of one stage in the life cycle of a bristlecone. Use labels and a caption to explain what you drew. Then work with your class to make a complete life cycle of a bristlecone pine.

## After reading Chapter 2, you will be able to:

- Describe animal life cycles and list characteristics that are common to all living things. **ANIMAL LIFE CYCLES**
- Compare and contrast complete and incomplete metamorphosis. **METAMORPHOSIS**
- Generalize the kinds of traits that are inherited and identify examples. **INHERITED TRAITS**
- Generalize the kinds of traits that are acquired and identify examples. **ACQUIRED TRAITS**
- Science in a Snap! Generalize the kinds of traits that are inherited and identify examples. **INHERITED TRAITS**

CHAPTER 2

# HOW DO ANIMALS AND

**Most animals** go through patterns of growth and change that are common to other kinds of animals. This baby sandhill crane is tiny with fuzzy feathers, but it already has other features like those of its parent. How do young grow and change to look like their parents?

Adult sandhill cranes in Florida can be as tall as most 10-year-old humans!

TECHTREK
myNGconnect.com

Student eEdition | Vocabulary Games | Digital Library | Enrichment Activities

# SCIENCE VOCABULARY

## metamorphosis
(met-ah-MOR-fah-sis)

**Metamorphosis** is a series of major changes in an animal's body form during its life cycle. (p. 54)

The adult stage is the last stage in metamorphosis.

## larva (LAR-va)

A **larva** is a young animal with a body form very different from the adult. (p. 55)

This wormlike larva will become an adult moth that flies.

## pupa (PŪ-pah)

A **pupa** is the stage in which the body form of a young animal changes from the larva to the adult. (p. 55)

An adult moth will come out of the pupa stage.

**heredity**
(hah-RED-ah-tē)

**larva**
(LAR-va)

**metamorphosis**
(met-ah-MOR-fah-sis)

**nymph**
(nimf)

**pupa**
(PŪ-pah)

Vocabulary Games

## nymph (nimf)

A **nymph** is the stage in which the young animal looks like the adult. (p. 56)

Unlike an adult praying mantis, the nymph does not have wings.

## heredity (hah-RED-ah-tē)

**Heredity** is the passing of traits from parents to their offspring. (p. 60)

Markings, such as stripes or spots, are part of an animal's heredity.

# Animal Life Cycles

Observe the cows in the photograph. Find the baby cow. Like all living things, cows reproduce more of their own kind. An animal's life cycle tells how the animal is born, grows, changes into an adult, and reproduces. In a cow's life cycle, the young are born alive and drink milk until they can eat other foods. They grow into adults and have young, repeating the life cycle.

TECHTREK
myNGconnect.com
Digital Library
Name some ways the calf looks like the parent.

Cows also share more characteristics with other living things. They take in nutrients, or food, and give off wastes. They take in oxygen from the air. Other living things take in oxygen from water, or take in other gases. Cows also get larger as they grow, just like other living things. A cow can live around 20 years before it dies.

**Cows share certain characteristics with other living things.**

A dolphin calf is born alive, but underwater! Soon after birth, the mother dolphin quickly pushes the calf to the water's surface for air. After three to six years, the calf will leave its mother to live on its own. As an adult, it will continue the life cycle of dolphins by having offspring.

The Atlantic spotted dolphin calf has the same body shape, fins, and pointed nose as its mother. What is different?

Unlike dolphins, loggerhead turtles hatch from eggs. When female loggerhead turtles are ready to lay eggs, they travel to the same beaches where they hatched. Once on the beach, the turtles dig small holes and lay their eggs inside. Soon the baby turtles hatch and make their way to the water. When the turtles become adults, the females return to the same beach and lay their eggs.

## LIFE CYCLE OF A **LOGGERHEAD TURTLE**

A female loggerhead turtle lays many eggs at once.

Young turtles hatch, or break out, from their eggs.

Adult turtles live in the ocean.

## Before You Move On

1. What are the stages in a loggerhead turtle's life cycle?
2. Compare a dolphin's life cycle to that of a cow.
3. **Apply** Name two other animals that have similar life cycles as that of the cow.

# Metamorphosis

Some kinds of animals, such as amphibians and insects, go through **metamorphosis** during their life cycles. Metamorphosis is a series of major changes in an animal's body form as it moves through its life cycle.

**Complete Metamorphosis** Insects with four life cycle stages undergo complete metamorphosis. Find the egg stage of the luna moth in the diagram. Trace the luna moth's life cycle with your finger.

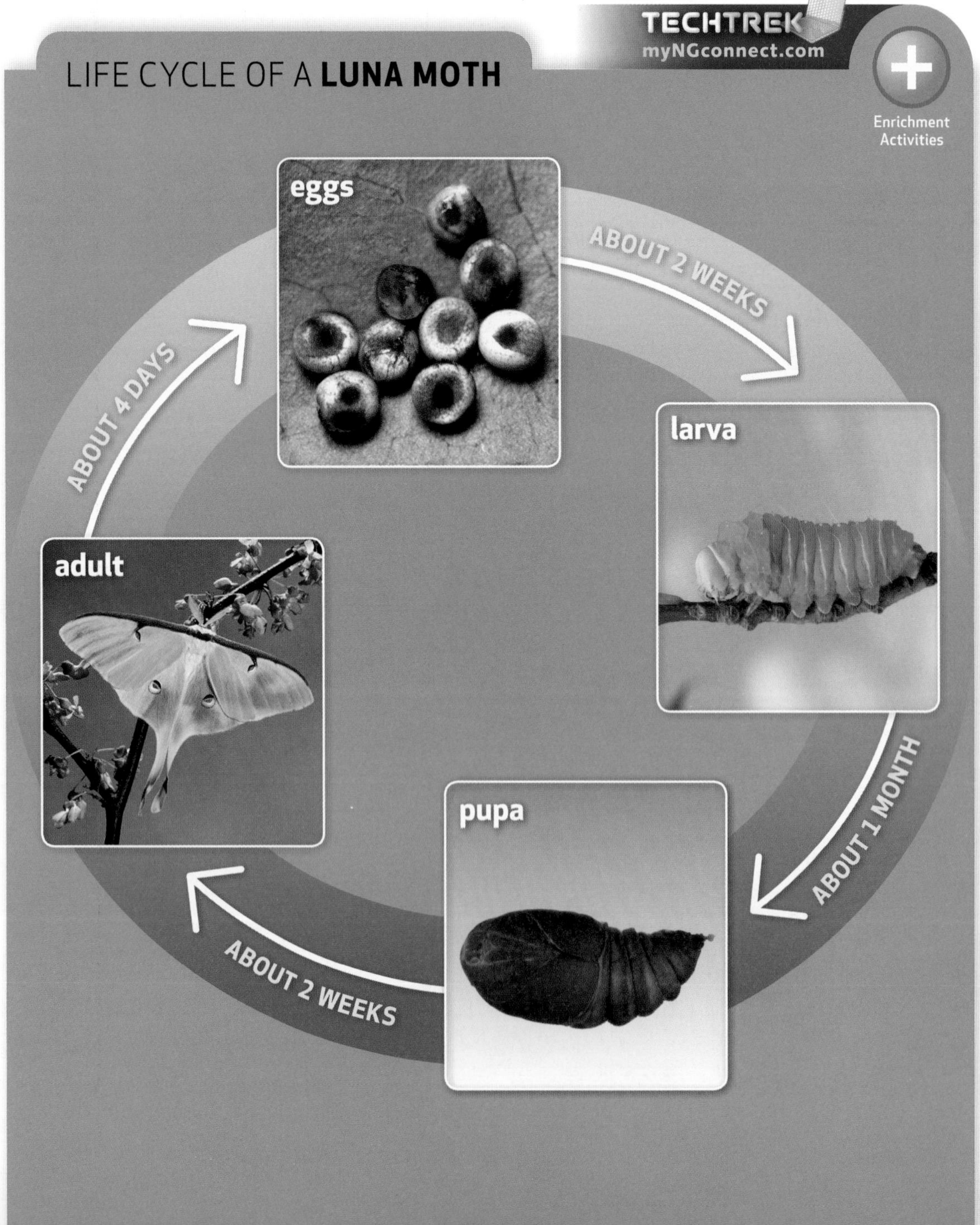

The tiny eggs hatch into caterpillars, which are the larvae. In the **larva** stage, the young insect is in a body form very different from the adult. Next, the insect enters the **pupa** stage. The pupa is the stage in which the body form changes from the larva to the adult. The luna moth pupa forms a cocoon, or hard case, around its body during this stage. When the metamorphosis is finished, the adult breaks out of the cocoon. In the adult stage, the luna moth lives just long enough to lay eggs.

This luna moth is in the adult stage of its life cycle.

**Incomplete Metamorphosis** Some animals, such as praying mantises, go through metamorphosis in a three-stage life cycle. This is called incomplete metamorphosis. The three stages are egg, **nymph**, and adult. A nymph is the stage of incomplete metamorphosis in which the young animal looks like the adult but may not have some features, such as wings. Praying mantis eggs soon hatch into the nymph stage. Over time, the nymph grows larger as it becomes an adult.

An adult praying mantis can lay hundreds of eggs.

The life cycle of a praying mantis is very different from the luna moth. One important difference is that during complete metamorphosis, a larva totally changes during the pupa stage. But in incomplete metamorphosis, the young praying mantis looks very much like the adult. The nymph grows larger, and its body form changes very little. But it does not grow wings until just before it changes into an adult.

Digital Library

## LIFE CYCLE OF A **PRAYING MANTIS**

eggs (in capsule)

ABOUT 3 WEEKS

nymph

ABOUT 6 MONTHS

adult

ABOUT 6 MONTHS

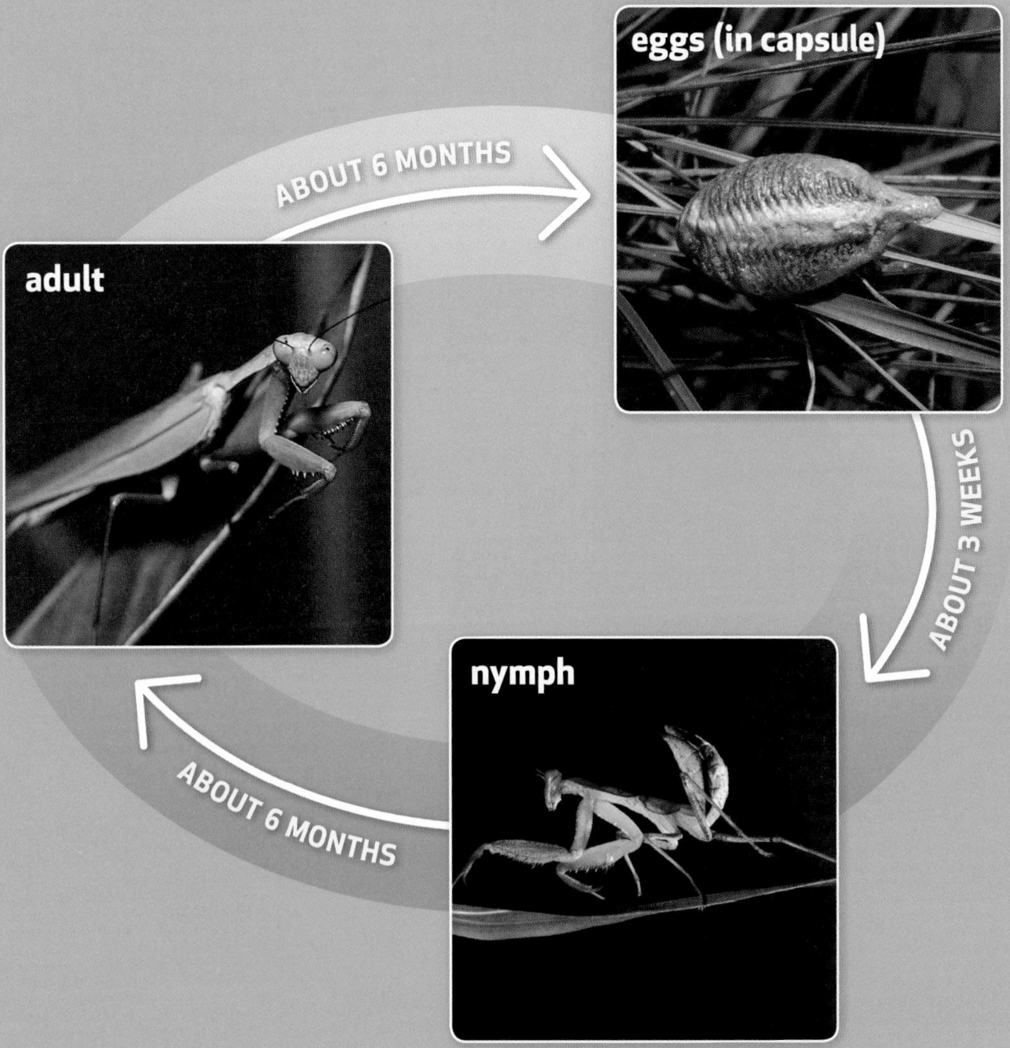

**Molting** As a nymph grows, its outer covering hardens. It becomes almost like a shell. As the nymph grows bigger, that hard outer covering becomes too small. The nymph molts, or sheds that covering. The nymph grows bigger and its covering hardens again. When this covering becomes too small, the nymph molts again. Many insects molt several times during the nymph stage.

**Dogday harvestfly cicada nymphs live underground. They crawl into trees when they molt for the last time.**

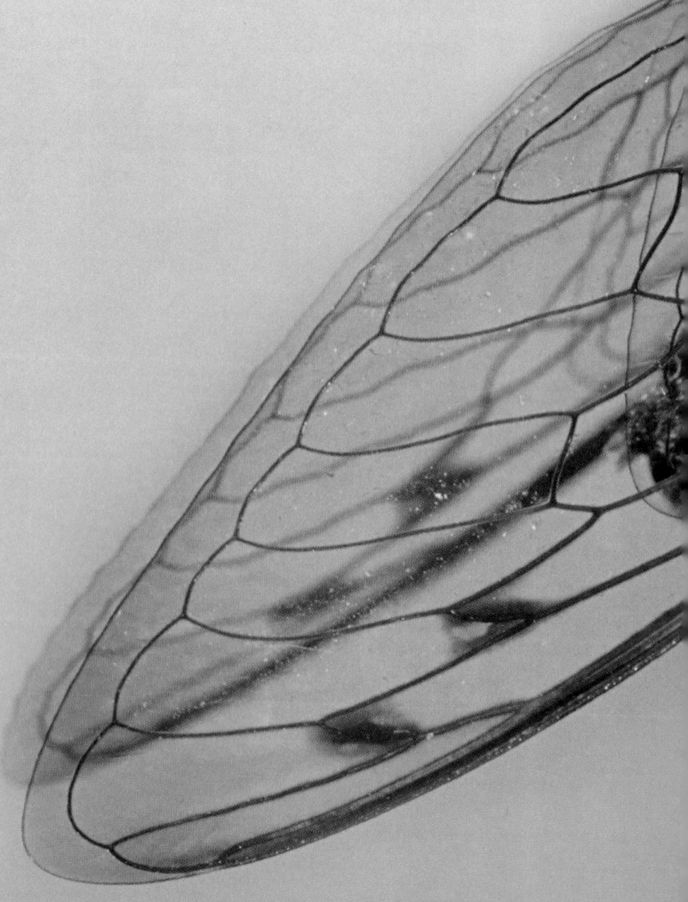

Finally, the insect goes through its last molt. If you see an insect with wings shedding its outer shell, it is most likely the adult. So if you see a tiny fly with wings, for example, that kind of fly is just tiny. It will not grow into a bigger fly.

**The adult female dogday harvestfly cicada cuts open a twig and lays eggs inside it. The eggs hatch in about six weeks.**

## Before You Move On

1. How does molting allow an animal to grow bigger?
2. How is complete metamorphosis different from incomplete metamorphosis?
3. **Generalize** How is a nymph different from a pupa?

# Inherited Traits

**Inherited Physical Traits** Look at the picture of the elephants below. How is the baby like its mother? You see that it is the same color, for example. An animal's color is caused by **heredity**. Heredity is the passing of traits from parents to their offspring. Skin color, body shape, and number of legs are some inherited traits. Offspring inherit the general traits of that particular kind of animal. They also inherit traits that make them special as individuals, such as the size and shape of the tusks.

**What traits did these young elephants inherit that tell you they are elephants?**

Many kinds of animals inherit traits from two parents. The offspring will be similar to both parents. But it will not be exactly like either parent. In American alligators, the size of the head, button-shape scales on the chest, and the roundness of the snout are some differences you can see. These traits will be different from one alligator to the next. An individual's special traits are caused by how the parents' traits work together.

**The baby alligator has stripes that fade as it grows. As an adult it will look much like its parents and other American alligators.**

## Science in a Snap! Colorful Characteristics

Find the adult in this photograph. Closely observe its traits. Then closely observe the traits of the young.

**Which of the traits you observed in the young do you think are inherited from the parents?**

**Inherited Behaviors** Many animal behaviors, or actions, are also inherited. For example, a bird inherits the ability to fly. More complex inherited behaviors are instincts. Instincts help animals find or catch food, protect themselves, and reproduce. A spider inherits its ability to make a certain kind of web. The kind of web a spider weaves is related to the kind of food it eats.

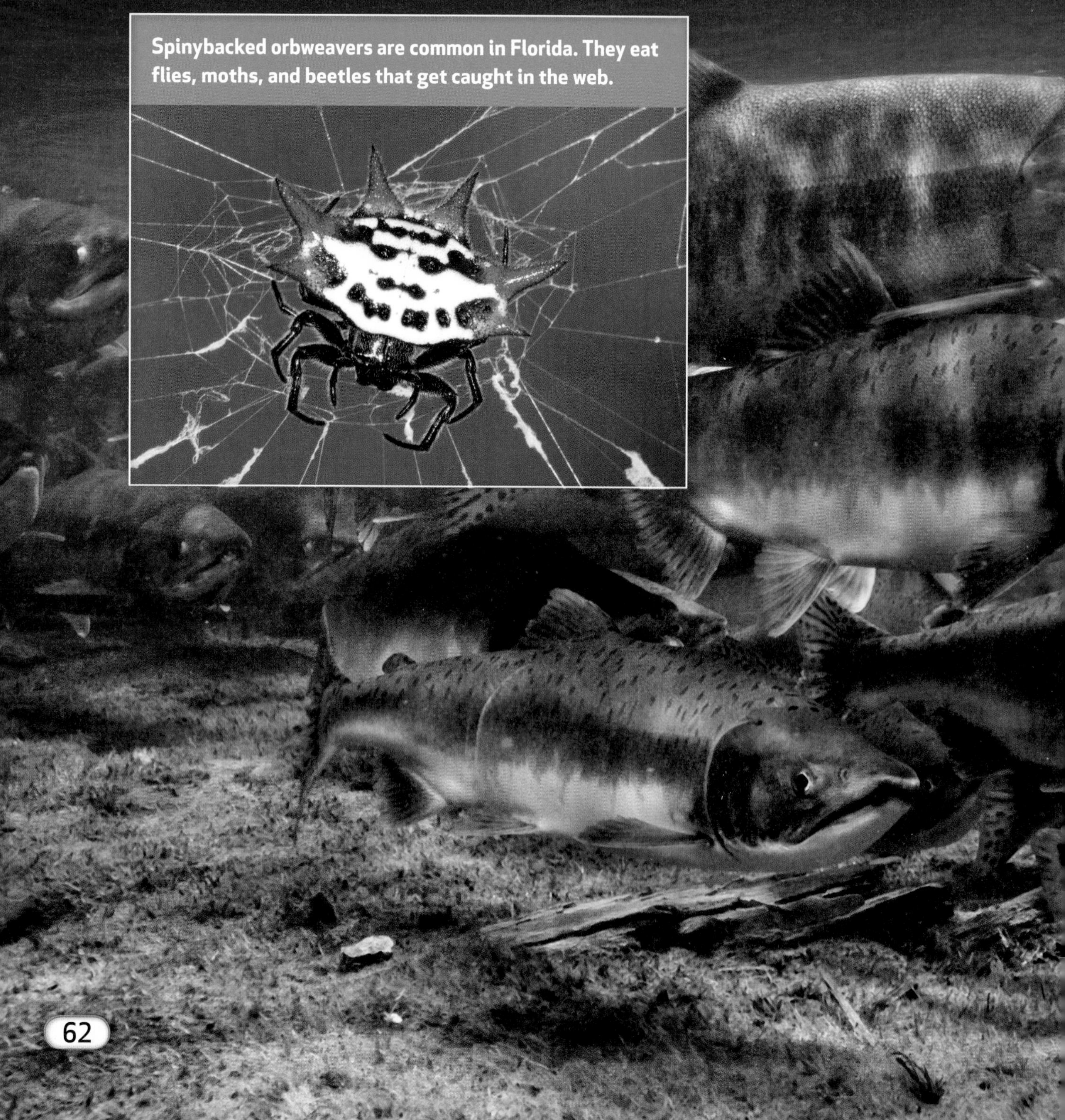

Spinybacked orbweavers are common in Florida. They eat flies, moths, and beetles that get caught in the web.

Animals follow their instincts all through their lives. For example, young salmon hatch in streams. Then they follow their instincts to swim to the ocean. Some salmon swim more than 1,600 kilometers (1,000 miles) in this journey to the ocean! Years later, as adults, they follow their instincts and return to the streams where they hatched. They swim upstream past dams, fishermen, and other hazards. Once they reach the stream where they hatched, the females lay their eggs.

These pink and chum salmon left Hartney Creek in Alaska when they were about the size of your finger. By instinct they have returned to lay eggs.

## Before You Move On

1. What is heredity? List some inherited traits.
2. How do instincts help an animal survive? Give an example.
3. **Draw Conclusions** A bird knows how to build a nest without ever learning how. Explain how this is possible.

# Acquired Traits

**Acquired Physical Traits** Not all traits are inherited. Some are acquired, or gained. These traits can be acquired because of behaviors or factors in the environment. The topi below might have acquired the trait of one horn during a fight over its territory.

**Having two horns is an inherited trait in topis. This topi has the acquired trait of one horn.**

Animals can also acquire physical traits from their diet. A diet is all the foods an animal eats. Diet affects an animal's body size, weight, and health. In some animals, diet can even change body color! Flamingos, for example, are born with white feathers. Flamingos have diets of algae, insect larvae, and shrimp. Substances in these foods cause a flamingo's feathers to turn pink. The more of these foods they eat, the pinker they get!

The colors of a flamingo's feathers can change from white to deep pink.

**Learned Traits** Some acquired traits result from learning. Many kinds of animals learn behaviors. Learning can be simply getting used to something happening. A squirrel in your backyard learns that you are not a predator. So it will allow you to stand closer to it than a squirrel in the wild would. The capuchin monkey below shows more complicated learning.

The capuchin finds a good place to put the palm nut.

The capuchin balances on its tail and gets ready to slam the rock down on the palm nut.

Young capuchin monkeys learn that they can crack open a nut using a heavy rock. Scientists have observed that the monkeys hold the rock in a certain way so they can hit the nut harder.

Adult birds such as cardinals and warblers learn the songs they sing to attract mates. Your pet dog or cat can learn where and when you will feed it. What other behaviors do you think a pet cat or dog can learn?

**Notice that the capuchin is still balancing itself with its tail as it slams the rock down. Did the palm nut break open? What do you think the capuchin will do next?**

## Before You Move On

1. What is an acquired trait? Give an example.
2. How do animals acquire traits?
3. **Draw Conclusions** Do you think it would be easy to train a capuchin to use a hammer to hit a nail? Explain why or why not.

NATIONAL GEOGRAPHIC

# ON THE JOB WITH SERVICE DOGS

Did you know that many dogs are trained to help people? They guide people who cannot see, and they help find missing people. To do these jobs, dogs use both instincts and learned behaviors.

TECHTREK
myNGconnect.com

Digital Library

**Border collies have a natural instinct for herding, which makes them easy to train to help people.**

Dog trainers often begin by teaching dogs to follow a few simple commands, such as "sit," "stay," and "heel." When the dogs obey the commands, the trainers may reward them with food treats. With practice, dogs can learn more difficult behaviors.

Guide dogs get very specific training to lead and protect their blind owners. Other dogs that work as police dogs receive training to use their speed, strength, and sense of smell. This helps the police do many different jobs.

**A guide dog learns how to help its owner walk through a park, use an escalator, or enter a bus or subway.**

**This golden retriever is being trained as a search-and-rescue dog. To rescue a person, the dog may need to walk across high, narrow paths.**

All animals go through life cycles, or stages of growing, changing, and reproducing. Many animals go through metamorphosis, a series of changes in body form. Young animals grow up to look much like their parents, but are not exactly the same as their parents. The young inherit traits and many behaviors from their parents. Other traits are acquired or learned.

**Big Idea** Animals inherit traits from their parents that determine how they grow, change, and reproduce.

## Vocabulary Review

**Match** each of the following terms with its correct definition.

**A. pupa**
**B. heredity**
**C. larva**
**D. metamorphosis**
**E. nymph**

1. The stage in which the body form of a young animal changes from the larva to the adult
2. A series of major changes in an animal's body form during its life cycle
3. The stage in which the young animal looks like the adult
4. The passing of traits from parents to their offspring
5. The stage of a young animal with a body form very different from the adult

## Big Idea Review

1. **Identify** What are the two kinds of traits that animals can have?

2. **Name** What are the three stages of incomplete metamorphosis?

3. **Compare and Contrast** How do the life cycles of dolphins and sea turtles compare?

4. **Cause and Effect** Dogs have an inherited trait for long tails. Some dogs, such as boxers, have a very short tail when they are adults. How did they most likely get this trait?

5. **Draw Conclusions** Why are instincts especially important for such animals as fish and turtles that leave their parents as soon as they hatch?

6. **Analyze** In a zoo, birds called oystercatchers are fed pieces of raw fish from the time they are born. In the wild, oystercatchers learn to break open clams and oysters for food. What do you think would happen if the oystercatchers from the zoo were released into the wild?

## Write About Metamorphosis

**Compare and Contrast** How are the larva and pupa stages in the life cycle of the luna moth different? How are they alike?

larva

pupa

CHAPTER 2

# LIFE SCIENCE EXPERT: LEPIDOPTERIST

## Karen Oberhauser

Would you like to be a lepidopterist? It's a person who studies butterflies! Karen Oberhauser is a lepidopterist who studies and teaches college classes about monarch butterflies. She also works with different groups to teach people how to conserve, or save and protect, monarch butterflies.

As a child, Oberhauser kept snakes. Now she studies butterflies!

**What did you like about science when you were in grade school?**

Most of my science memories are about things I did with my family. We went camping, went exploring, and helped my dad, who was a large-animal veterinarian, or doctor.

One reason many volunteers help Oberhauser collect data is because monarchs are easy to identify.

**What would you say has been the most interesting part of your job?**

I loved having a job that I could share with my daughters as they were growing up. When you study something such as monarch butterflies, it's fun for kids to be involved.

**Have there been any big surprises in your job?**

I've studied many features of monarch butterflies, from how they reproduce to how they migrate. It's been fun to become an expert on this butterfly and how it fits into its environment. I've also loved being a part of teaching people how important it is to preserve the natural world for the future.

**What has been your greatest accomplishment in your job?**

I'm proud that I have worked hard to plan programs that conserve habitats of monarch butterflies and other animals. I have worked to get people all over the United States to take part in collecting data about monarch butterflies. My goal has been to conserve as much of the natural world as possible. I feel that one way I have done that is by helping people see the wonder of the natural environments in this world.

Monarch wings are made of thousands of tiny overlapping scales, like shingles on a roof.

NATIONAL GEOGRAPHIC

# BECOME AN EXPERT

## Nature's Transformers

Many kinds of animals are transformers! They change during different stages in their lives. As young, they can look completely different than they do as adults. This metamorphosis occurs not just in insects such as beetles and dragonflies, but in other kinds of animals, too. Frogs, sea stars, and lobsters are some other kinds of animals that go through metamorphosis.

Identify some of the features of this young squat lobster.

**metamorphosis**

**Metamorphosis** is a series of major changes in an animal's body form during its life cycle.

Student eEdition

Digital Library

Squat lobsters live in the ocean in many places around the world. Look at how the squat lobster changes as it grows from a young lobster to an adult. Find a feature that both forms have, such as claws. Then tell how that feature is different in the two forms.

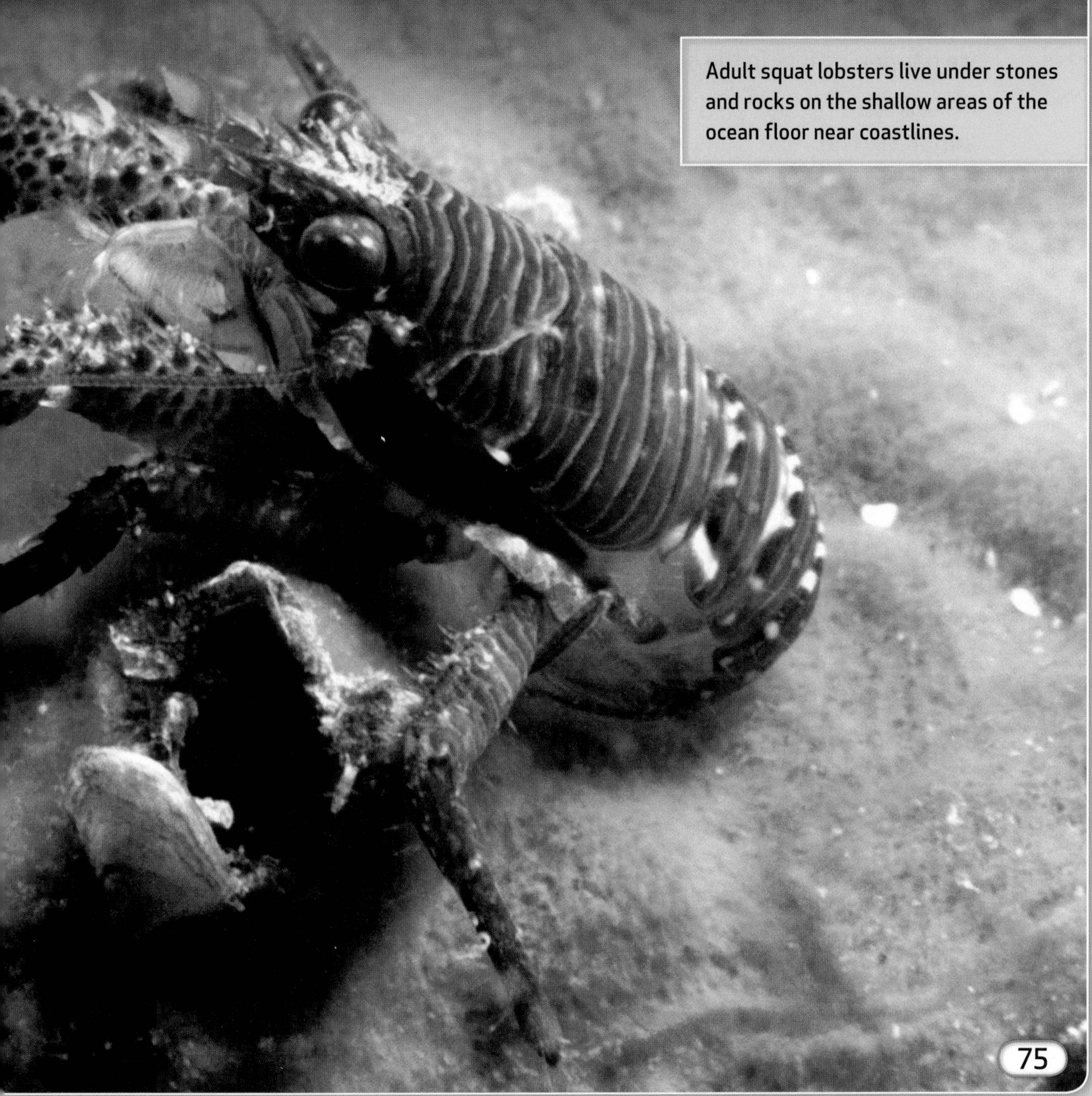

Adult squat lobsters live under stones and rocks on the shallow areas of the ocean floor near coastlines.

# From Grubs to Beetles

This adult rhinoceros beetle looks pretty scary. It's about 5 centimeters (2 inches) long! A rhinoceros beetle is an animal that goes through complete metamorphosis. Its life cycle takes about 18 months. The egg hatches and a **larva** forms. After many months, a hard shell grows around the larva's body. This is the **pupa** stage. Big changes are happening inside the beetle pupa. The larva dissolves. New body parts grow. The horn begins to form. After a few months, the change is complete! The pupa transforms into an adult beetle.

Rhinoceros beetles are strong. They can lift 850 times their own weight!

**larva**

A **larva** is a young animal with a body form very different from the adult.

**pupa**

A **pupa** is the stage in which the body form of a young animal changes from the larva to the adult.

# LIFE CYCLE OF A RHINOCEROS BEETLE

An adult rhinoceros beetle lives about 5 months.

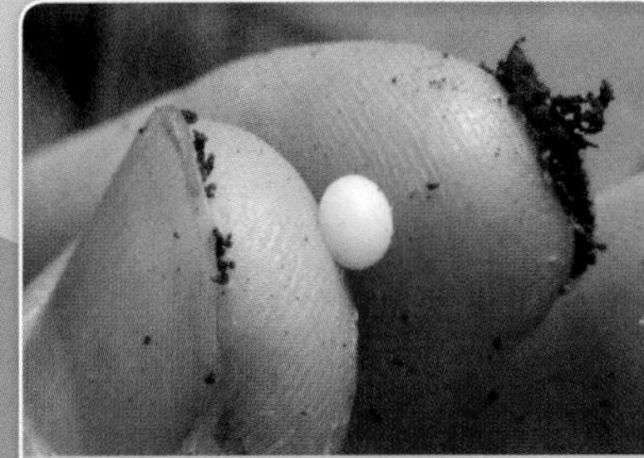
Each egg is about twice as big as the head of a pin.

In the larva stage, it grows from a small brown grub to a larger white form.

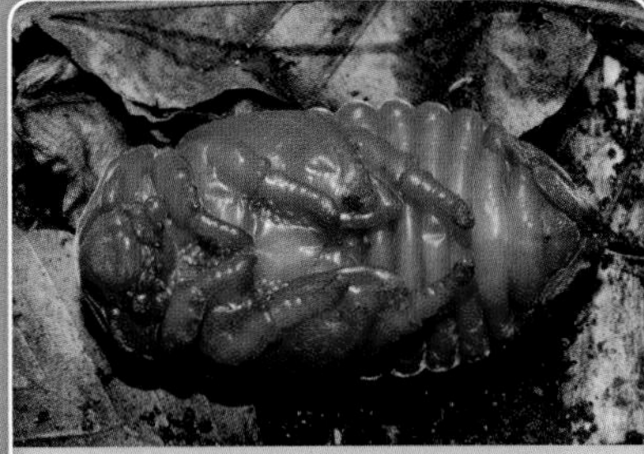
Within the pupa, the larva changes to an adult.

# From Nymphs to Dragonflies

Some insects, such as this dragonfly, go through incomplete metamorphosis. In this life cycle, there is no larva or pupa stage. A young dragonfly, or **nymph**, looks much like an adult, but it doesn't have wings. A nymph has a hard outer covering. As the nymph grows, it molts, or sheds, its covering. Some nymphs can live up to five years before they change into adults!

Metamorphosis is nearly complete. Soon a winged four-spotted chaser dragonfly will break out of its hard outer covering.

**nymph**

A **nymph** is the stage in which the young animal looks like the adult.

Notice how much longer the adult dragonfly is than the hard covering it just broke out of!

# From Tadpoles to Frogs

Frogs are another animal that go through complete metamorphosis. Find the eggs in the frog life cycle diagram. The eggs hatch into the larva stage—tiny tadpoles. Tadpoles must breathe oxygen like all other animals. But to do so, they must breathe through gills to get oxygen from the water.

Eggs take anywhere from 6 days to 3 weeks to hatch into larvae.

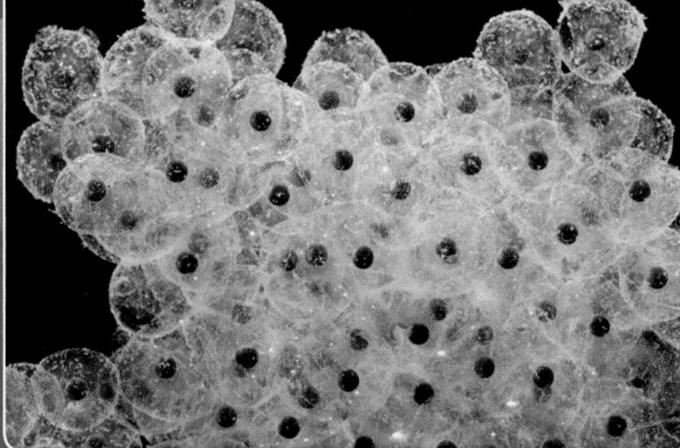

An adult frog has lungs and has made the transformation to living on land.

As the tadpoles grow, they begin to change. Tiny feet and legs form. Arms begin to stick out, elbows first! Their heads grow and eyes bulge. After about 12 weeks, tadpoles change into tiny frogs with tails. The frogs now breathe through lungs and can live on land. Soon their tails disappear, and complete metamorphosis has taken place.

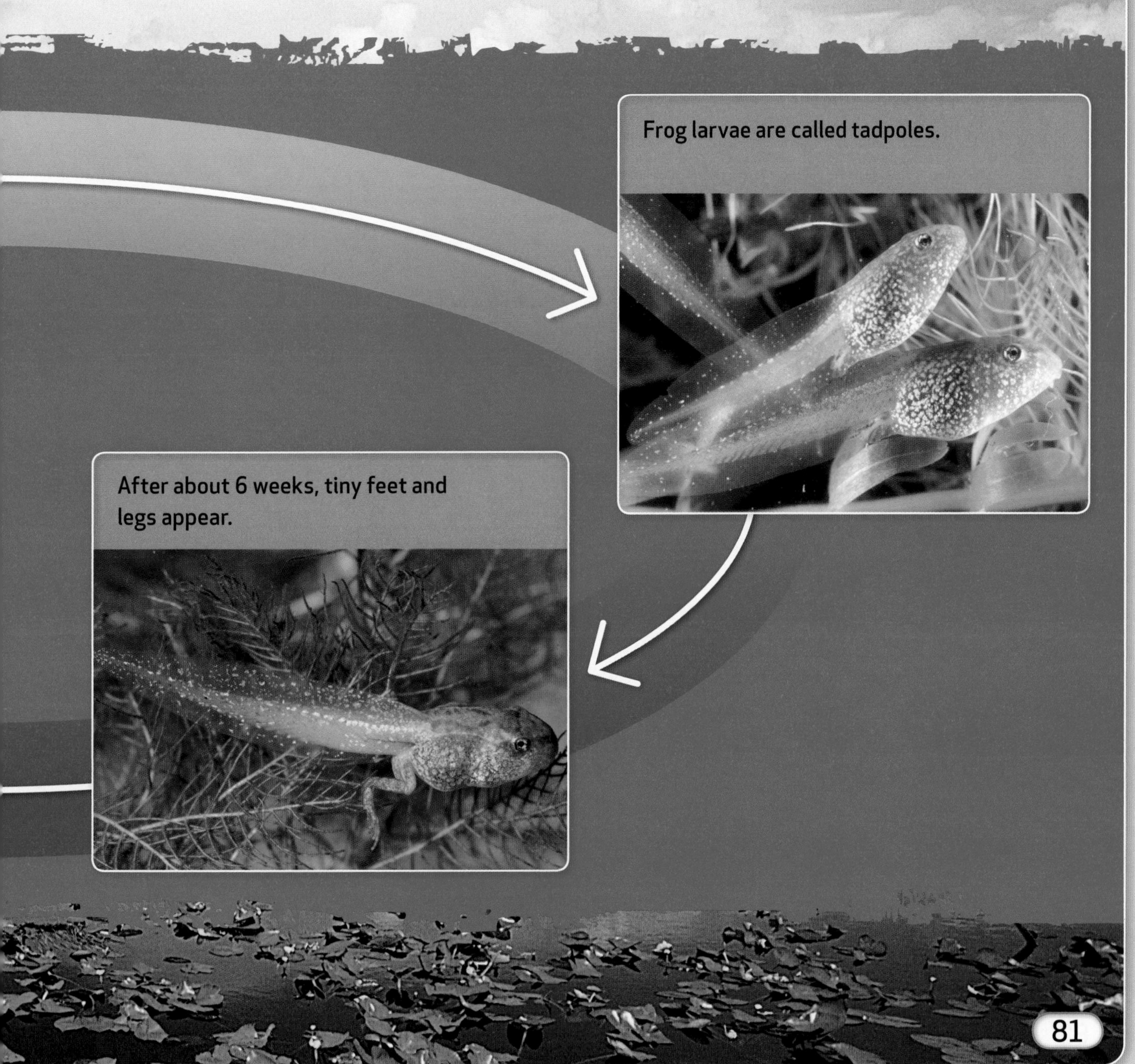

Frog larvae are called tadpoles.

After about 6 weeks, tiny feet and legs appear.

# Colorful Frogs

You might think that all frogs are green, but that is not so! Take a look at the different kinds of frogs on these pages. How does a blue poison dart frog become blue? Does a red tomato frog look like a tomato or its parents? Frogs get their traits through **heredity**, the passing of traits from parents to their offspring. So the blue poison dart frogs are blue because their parents were blue. The tomato frog gets its tomato-like appearance from its parents.

Blue poison dart frogs give off poison from their skin. This protection keeps them from being eaten by other animals.

**heredity**

**Heredity** is the passing of traits from parents to their offspring.

Does this mean that all blue poison dart frogs look exactly alike? No. Each individual will have slightly different black markings on its back. If you looked closely, you could tell one from all the rest. Traits such as skin color, number of limbs, and size are all inherited. A frog may look very much like its parents, but offspring are all unique, or one of a kind.

TECHTREK
myNGconnect.com
Digital Library

This larva is beginning to show some traits of the adult.

Markings on these adult poison dart frogs can be green, blue, or yellow.

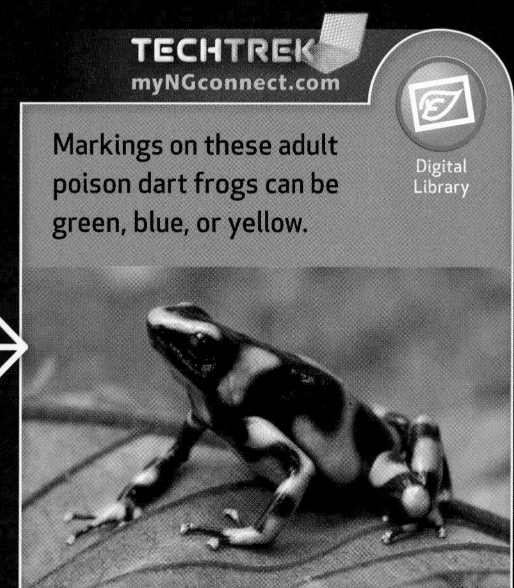

Tomato frogs are found only in Madagascar, a country off the coast of Africa.

NATIONAL GEOGRAPHIC

CHAPTER 2

# SHARE AND COMPARE

**Turn and Talk** How can animals grow, change, and reproduce? Form a complete answer to this question together with a partner.

**Read** Select two pages in this section. Practice reading the pages. Then read them aloud to a partner or a small group. Talk about why the pages are interesting.

my SCIENCE notebook **Write** Write a conclusion that tells the important idea you learned about metamorphosis. State what you think is the Big Idea of this section. Share what you wrote with a classmate. Compare your conclusions.

my SCIENCE notebook **Draw** Work in small groups of three or four. Choose one of the animals you learned about. Have each person draw a diagram of one stage in its life cycle and then label it. Put the diagrams together in a life cycle and write captions for it as a group. Compare your life cycle with those of other groups.

CHAPTER 3

HOW DO LIVING THINGS DEPEND ON THEIR ENVIRONMENT?

TECHTREK
myNGconnect.com

**The African savanna** is rich in life. The cape buffalo, cattle egret, and grasses shown here are part of this richness. These living things depend on each other. Cape buffalo graze on the grasses. Cattle egrets and other birds find food by living closely with cape buffalo and other grazers. All of these living things depend on the sun. How?

This cape buffalo and cattle egret depend on each other and the rest of their environment.

86

87

## After reading Chapter 3, you will be able to:

- Recognize and explain how animals and plants get the energy they need for life. **WHAT LIVING THINGS NEED FROM THEIR ENVIRONMENT**
- Classify living things as producers or different kinds of consumers. **COMMUNITIES OF LIVING THINGS**
- Distinguish among herbivores, carnivores, and omnivores, and give examples. **WHAT LIVING THINGS NEED FROM THEIR ENVIRONMENT**
- Describe how energy passes from one living thing to another in a community. **FOOD CHAINS AND FOOD WEBS IN COMMUNITIES**
- Identify the role that each organism in a food web plays. **FOOD CHAINS AND FOOD WEBS IN COMMUNITIES**
- Explain how living things depend on and compete with each other. **INTERACTIONS AMONG LIVING THINGS**
- Describe forest, prairie, tundra, pond, and ocean environments and the types of plants and animals they support. **DIFFERENT ENVIRONMENTS**
- Science in a Snap! Describe how energy passes from one living thing to another in a community. **FOOD CHAINS AND FOOD WEBS IN COMMUNITIES**

CHAPTER

# 3

# HOW DO LIVING THINGS DEPEND ENVIRON

**The African savanna** is rich in life. The cape buffalo, cattle egret, and grasses shown here are part of this richness. These living things depend on each other. Cape buffalo graze on the grasses. Cattle egrets and other birds find food by living closely with cape buffalo and other grazers. All of these living things depend on the sun. How?

This cape buffalo and cattle egret depend on each other and the rest of their environment.

**TECHTREK**
**myNGconnect.com**

Student eEdition

Vocabulary Games

Digital Library

Enrichment Activities

# SCIENCE VOCABULARY

## herbivore (UR-buh-vor)

An **herbivore** is an animal that eats plants to survive. (p. 94)

A cow is an herbivore because it eats grass and other plants.

## carnivore (KAR-nuh-vor)

A **carnivore** is an animal that eats other animals to survive. (p. 95)

This leopard is a carnivore because it kills and eats other animals.

**carnivore**
(KAR-nuh-vor)

**herbivore**
(UR-buh-vor)

**omnivore**
(AHM-ni-vor)

**predator**
(PRED-uh-tur)

**prey**
(PRĀ)

## omnivore (AHM-ni-vor)

An **omnivore** is an animal that eats plants and other animals to survive. (p. 96)

Omnivores, such as baboons, can eat both plants and animals.

## predator (PRED-uh-tur)

A **predator** is an animal that hunts other animals for food. (p. 98)

A snow leopard is a predator because it hunts and eats other animals for food.

## prey (PRĀ)

**Prey** are animals that other animals hunt for food. (p. 98)

The crayfish is prey for the heron.

# What Living Things Need from Their Environment

Find the living things in this picture. Each kind of living thing is different from the others. But like you, they all need air, water, energy, and space to live. Their environment supplies all of these things. Plants also need soil. It provides some materials plants need to grow.

Living things need each other. One reason they need each other is to get energy from food. Plants use energy directly from the sun to make their own food. But animals can't use sunlight this way. They have to eat plants or other animals to get the energy they need to live and grow. What do you eat that comes from plants or animals?

## Before You Move On

1. What do all living things need?
2. How do animals get the energy they need?
3. **Predict** Suppose the lake shown here dried up. How would that affect the animals?

# Communities of Living Things

The plants and animals shown here have something in common. They belong to the same community. A community is all the different kinds of living things in an area. Members of a community interact with and depend on each other. For example, the panda interacts with the bamboo by eating it. The snow leopard interacts with the panda and other animals by eating them. These interactions provide energy for the animals. But how does the bamboo get its energy?

Giant panda bears eat almost nothing but bamboo.

This snow leopard depends on other animals for food. It has its eye on some food right now.

**Producers** Bamboo is a producer. So are the grass, trees, and other plants where you live. Producers use air, water, and energy from the sun to make their own food. They store this food in their leaves, stems, and other green parts. The food is a source of energy for the producers. It's also a source of energy for living things that don't produce their own food, such as pandas and snow leopards.

Bamboo is a kind of grass. Food is stored in its green leaves and stems.

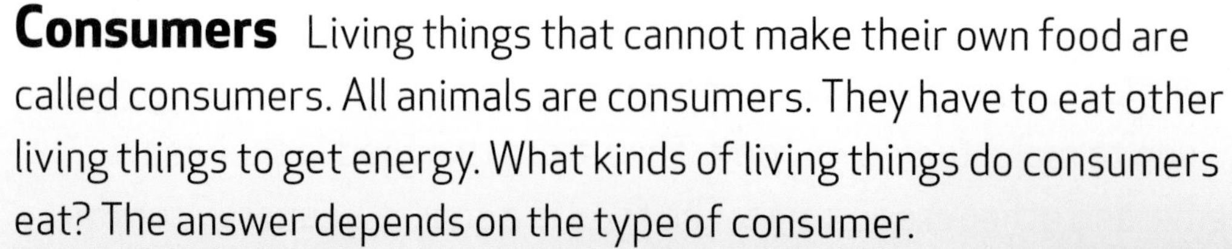

**Consumers** Living things that cannot make their own food are called consumers. All animals are consumers. They have to eat other living things to get energy. What kinds of living things do consumers eat? The answer depends on the type of consumer.

Cows, horses, goats, and sheep eat plants. An animal that eats plants is an **herbivore**. Large, flat teeth help herbivores grind leaves and stems.

**When you eat beef or drink milk, you get some of the energy the cows got by eating grass.**

Cats, seals, snakes, and hawks eat other animals. An animal that eats other animals is a **carnivore**. Most carnivores have sharp teeth or beaks that can bite and tear meat.

A leopard is a carnivore. This one guards the impala it killed for food.

Raccoons, skunks, pigs, and bears eat plants as well as other animals. An animal that eats both plants and other animals is an **omnivore**. Humans are omnivores too. Omnivores have some teeth that help with chewing plants and some teeth that bite and tear meat.

Eating certain kinds of foods is one of the roles, or jobs, of a consumer. The different roles help keep enough food for the community. For example, if all animals were herbivores, an area might soon have no plants left to eat.

This baboon is eating the meat of a young impala.

Baboons round out their diet with fruit.

**Decomposers** Consumers that feed on the remains of plants and animals are decomposers. Think of leaves that fall to the ground. After a few months, the leaves might have many holes. Decomposers are at work. As they break down the dead matter, they release some of the materials in that matter back into the soil. Then plants can use these materials to grow. Some decomposers are too small to see with your eyes alone.

TECHTREK
myNGconnect.com
Digital Library

Mushrooms, a type of decomposer, grow as they help decompose this forest log.

## Before You Move On

1. How do producers get food?
2. How are herbivores, carnivores, and omnivores alike? How are they different?
3. **Infer** Why are decomposers sometimes called "nature's recyclers"?

# Food Chains and Food Webs in Communities

When you see a squirrel eating a nut or the sun shining on a plant, you are seeing a link in a food chain. A food chain is one path that energy takes through a community.

Trace the flow of energy in the food chain shown here. Cattails and other producers use the sun's energy to make food. Consumers, such as crayfish, get some of that energy when they eat the cattails. Later, a great blue heron might grab the crayfish and eat it. The heron is a **predator** —an animal that hunts and eats other animals. The crayfish is the **prey** —an animal that is hunted and then eaten.

## **FOOD CHAIN** IN A POND

**Most food chains begin with the sun. Producers, such as cattails, use sunlight to make food.**

**Consumers, such as crayfish, eat the cattails and get the plants' stored energy.**

**Other consumers, such as this heron, eat the crayfish and get the energy stored in its body.**

## Science in a Snap! What's Your Food Chain?

Think about your favorite meal. Infer what plant or animal provided each kind of food. Include what the animals ate.

Draw and label each plant or animal on a card. Arrange the cards to show different food chains.

**Where do you fit into each food chain?**

**Tortoises get their energy from dandelions and other plants. Where would you place this photo in the food chain shown on the opposite page?**

**Food Webs** You eat more than one kind of food. Most other consumers do too. A heron's prey includes fish, insects, mice, snakes, and other animals. A raccoon eats just about anything it can get its paws on, from berries and nuts to worms and fish. So a community includes many food chains that are all linked to each other. All of these connected food chains make up a food web. The drawing shows part of a pond food web. What does the heron eat? What eats the fish?

TECHTREK
myNGconnect.com

Enrichment Activities

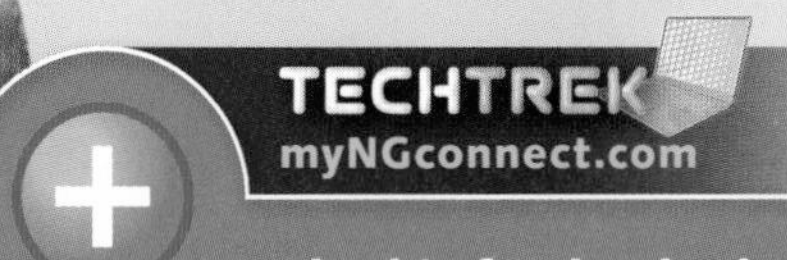

**In this food web, the arrows point from the thing being eaten to the thing that eats it. How many food chains can you find in this food web?**

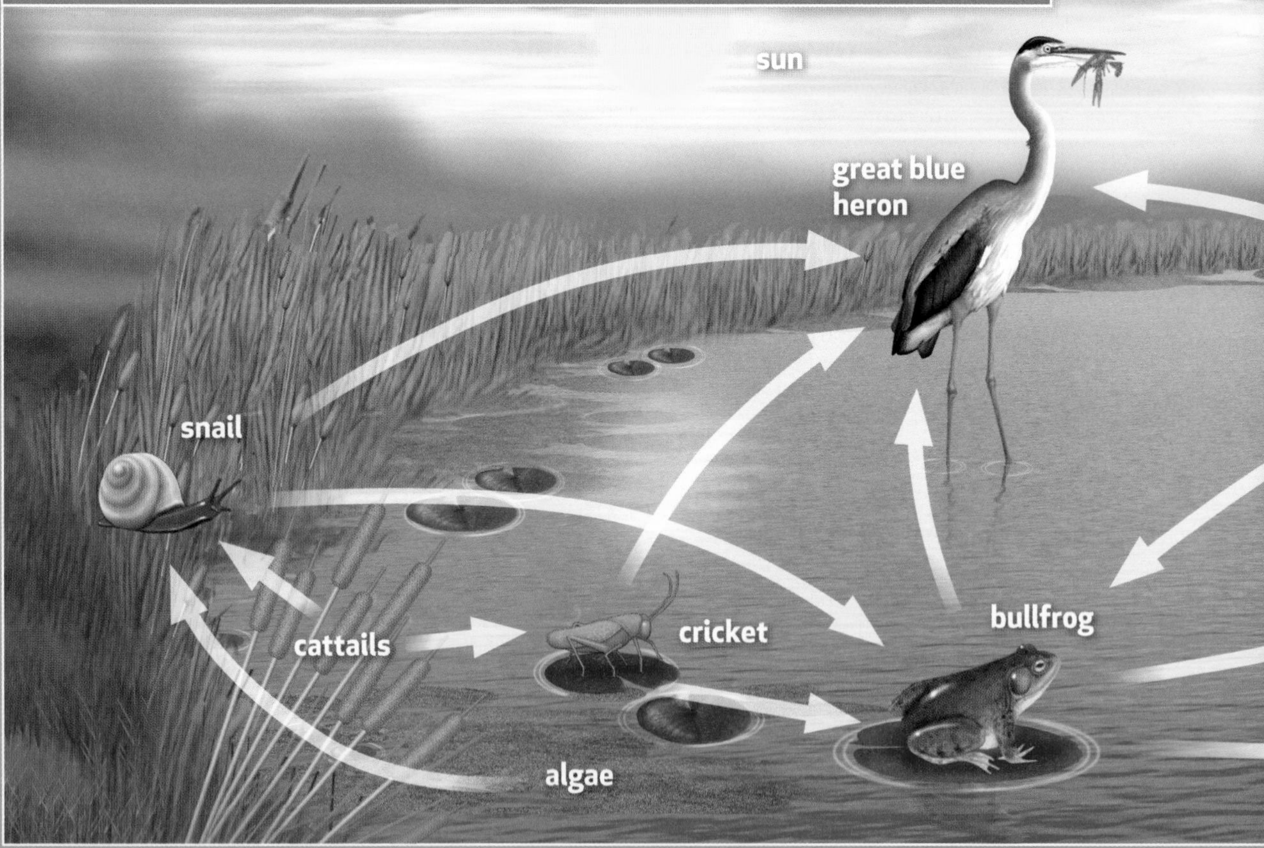

sun
great blue heron
snail
cattails
cricket
bullfrog
algae

What would happen if part of the food web changed? Suppose a disease kills most of the frogs. Animals that eat the frogs, such as herons and raccoons, would have less food. They might become weaker and not live as long. Over a few years, the numbers of these animals might decrease. Meanwhile, what would happen to the insects and other living things that the frogs eat? Their numbers would increase. One change in a food web can affect the whole community.

## Before You Move On

1. What is a food chain?
2. How is a food web different from a food chain?
3. **Predict** Suppose half the pond fills in with soil and grasses. How might that change affect the whole community?

NATIONAL GEOGRAPHIC

# WOLVES CHOOSE FISHING OVER HUNTING

Wolves are large predators. They chase down deer, moose, and other large prey. But scientists have discovered something surprising. When given the choice, wolves would rather fish than hunt.

Scientists have studied wolves in British Columbia, Canada. During most of the year, wolves eat deer and other land animals. But during the fall, thousands of salmon swim upstream to lay their eggs. That's when salmon become the wolves' favorite food.

**This wolf dives into the stream to catch a meal.**

Hunting salmon instead of deer makes a lot of sense. Wolves often get hurt while chasing down deer. The wolf might get gouged by antlers or kicked by powerful legs and sturdy hooves. Plus, the wolf has to use a lot of energy to chase the deer, and sometimes the deer gets away. It's a lot easier to just stand by the riverside and grab a fish as it swims by.

Wolves eat only the head of the salmon. Birds, crayfish, and other animals eat the body. Decomposers break down what's left. So the wolves' choice to go fishing supplies energy for much of the community.

**Wolves play an important role in food chains and food webs.**

# Interactions Among Living Things

Every time you talk with your friends, walk on the grass, or eat, you interact with living things. Living things interact with one another in different ways. These lion cubs interact with each other as they play. A bird interacts with a tree by building a nest in it. One of the most important ways living things interact is by getting food.

These cubs are learning important hunting skills as they play.

This lizard and beetle show how predators and prey interact. Predators need their prey for food. But what would happen if more and more lizards moved into the area? They may eat all the beetles. Then the lizards will have less food and may die.

The prey also need their predators. If fewer lizards are in the area, more and more beetles may live. But there may not be enough food, water, or space for all of them. Many prey animals may die of hunger or thirst. Predators and prey need each other to keep the community healthy.

This European green lizard begins to munch on a beetle.

**Parasites** When a bird lives in a nest in a tree, the bird does not harm the tree. But some consumers do harm other living things by living on or in them. These consumers are called parasites. They take nutrients from other living things. The animal or plant that provides the nutrients is called a host. Parasites harm their hosts but usually do not kill them.

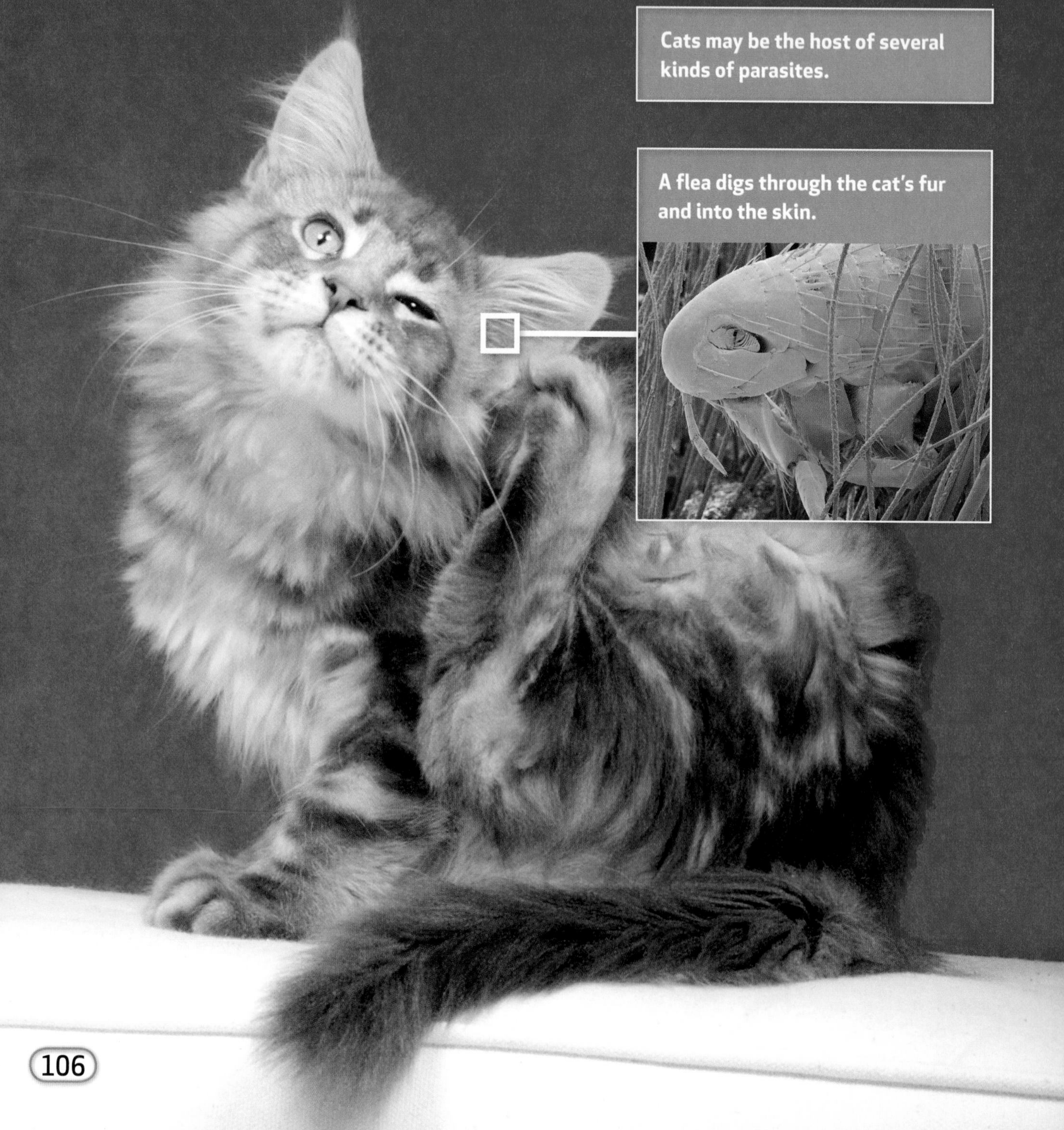

Cats may be the host of several kinds of parasites.

A flea digs through the cat's fur and into the skin.

Some parasites live on their hosts. Fleas dig through the skin of pets and other animals to suck their blood. After the meal of blood, the flea jumps off. Ticks, on the other hand, attach themselves to their hosts as they dig through the skin to suck blood.

Not all parasites are consumers. Some are producers. Look at the green clumps on the tree. These are plants called mistletoe. They take water and nutrients that travel through the tree trunk and branches.

These clumps of mistletoe are not part of the tree. They are parasites that live on the tree.

**Living Things Help Each Other** Sometimes living things help each other when they interact. For example, the bee in the photograph is helping the plant. The plant is also helping the bee. The flower of the plant makes yellow grains of pollen. The bee uses the pollen for food. The pollen sticks to the bee as it flies from flower to flower. The pollen then rubs off onto the other flowers. The flowers use the pollen to grow new plants.

**Bees carry yellow grains of pollen from flower to flower.**

Living things help each other in many other ways. In each interaction below, think about how both living things are helped.

## **HELPFUL** INTERACTIONS

**SPREADING SEEDS** Animals that eat the fruits of plants spread the plants' seeds to a new area. The plants may grow well in the new area.

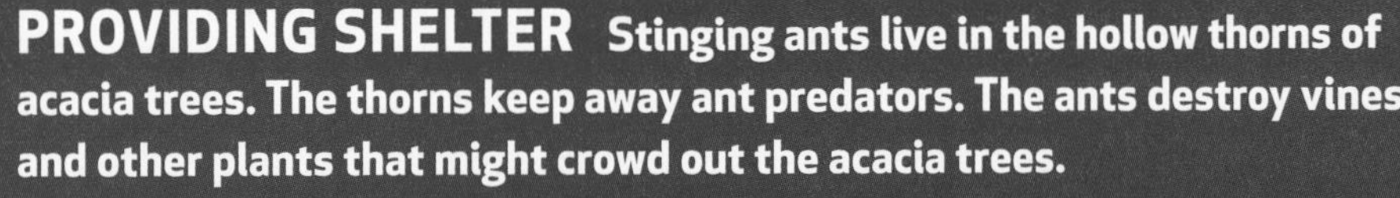

**PROVIDING SHELTER** Stinging ants live in the hollow thorns of acacia trees. The thorns keep away ant predators. The ants destroy vines and other plants that might crowd out the acacia trees.

**Living Things Compete with Each Other** If two hungry robins see the same worm, what do they do? They compete for it. The faster or stronger robin might end up with the worm. All living things need the same resources, such as food, water, and shelter. How are these horses competing with each other?

**These horses fight for water near a water hole.**

An area has only a certain amount of resources. Living things have to compete for them, like the robins trying to get the worm and the horses trying to get water. Now look at the picture below. How are the vines competing with the tree?

Strangler figs are plants that compete with trees for sunlight by growing up the trunks of the trees.

## Before You Move On

1. What is a parasite?
2. How do some bees and plants help each other?
3. **Infer** Why would a parasite benefit from its host being healthy?

# Different Environments

What different kinds of environments have you visited or seen in movies? Maybe you saw a cool, damp forest, a wide, windy prairie, or a cold, frozen tundra. Each of these places is an ecosystem. A large ecosystem has smaller ecosystems within it, such as a creek ecosystem or a pond ecosystem. Each kind of environment contains living and nonliving things that interact in that area. Compare the different environments on the next several pages.

**Eagles build nests high in the trees. They capture fish in streams and small animals on the ground.**

**A white-tailed deer is one of the many herbivores in this forest community.**

**A Forest Environment** Trees are the main plants in a forest. Smaller shrubs, ferns, mosses, and wildflowers grow on the ground under the trees. Different kinds of animals look for food and shelter in the different levels of a forest. Eagles build their nests high in the treetops. Black bear cubs may climb the lower branches to escape wolves or other dangers. Chipmunks live in burrows, or holes, in the ground. Look at the chart to see what other living things you might find in a forest. Think about what role each plays in the community.

## **FOREST** COMMUNITY

| | |
|---|---|
| **PRODUCERS** | Pine, spruce, fir, hemlock, oak, aspen, ferns, mosses |
| **CONSUMERS: HERBIVORES** | Deer, elk, moose, squirrels, chipmunks, beavers, termites |
| **CONSUMERS: CARNIVORES** | Bobcats, mountain lions, wolves, otter, mink, eagles, owls, hawks, salamanders, frogs |
| **CONSUMERS: OMNIVORES** | Brown bears, black bears, raccoons, foxes, most birds |
| **DECOMPOSERS** | Mushrooms |

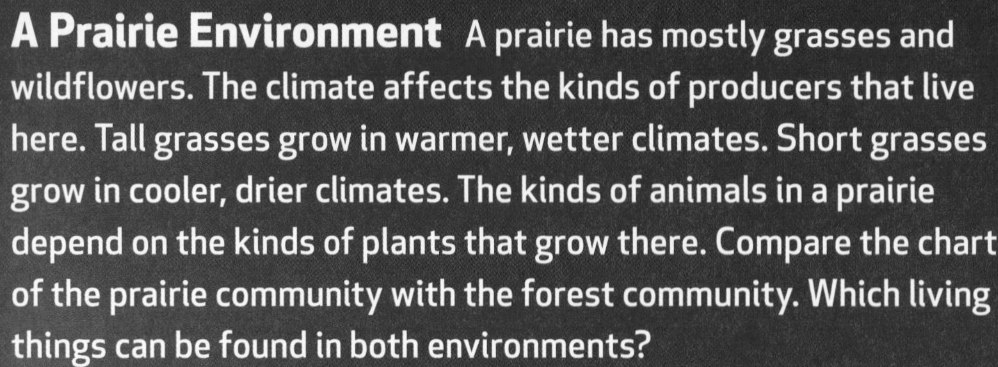

**A Prairie Environment** A prairie has mostly grasses and wildflowers. The climate affects the kinds of producers that live here. Tall grasses grow in warmer, wetter climates. Short grasses grow in cooler, drier climates. The kinds of animals in a prairie depend on the kinds of plants that grow there. Compare the chart of the prairie community with the forest community. Which living things can be found in both environments?

Black-tailed prairie dogs live in burrows on the short-grass prairies of the midwestern United States.

The purple flowers of the prairie blazing star provide food for bees, moths, and butterflies.

Field mice make their homes in the grasses of the prairie.

## PRAIRIE COMMUNITY

| | |
|---|---|
| PRODUCERS | Tall grasses, bunch grasses, clovers, paintbrushes, milkweeds, coneflowers, sunflowers, lupines |
| CONSUMERS: HERBIVORES | Deer, bison, prairie dogs, rabbits, gophers, mice, grasshoppers, bees |
| CONSUMERS: CARNIVORES | Coyotes, ferrets, eagles, owls, hawks, egrets, prairie chickens, frogs, dragonflies |
| CONSUMERS: OMNIVORES | Raccoons, badgers, skunks, rats, foxes, geese, ducks, most birds |
| DECOMPOSERS | Mushrooms |

**A Tundra Environment** Tundra environments are found near the Arctic and high in the mountains. The tundra is one of the harshest environments on Earth. The climate is very cold for most of the year. The ground is frozen throughout most of the year. As a result, few large plants grow there. Instead, the ground is covered with lichens, mosses, and small flowering plants. Lichens often form colorful patches on rocks and soil.

Denali National Park in Alaska is tundra country.

The tundra is, however, home to many large animals. Some, such as caribou, moose, and musk oxen, are herbivores. Others, such as polar bears, are carnivores.

**Polar bears feed on seals and fish during summer to build up a thick layer of fat. The fat keeps them warm in winter.**

**Shoots of small trees grow among the white lichens and blueberry plants in Alaska.**

## **TUNDRA** COMMUNITY

| | |
|---|---|
| **PRODUCERS** | Lichens, mosses, ferns, grasses, willows, lupines |
| **CONSUMERS: HERBIVORES** | Caribou, musk oxen, chipmunks, rabbits |
| **CONSUMERS: CARNIVORES** | Polar bears, wolves, otter, mink, eagles, snowy owls, hawks, salamanders, frogs |
| **CONSUMERS: OMNIVORES** | Brown bears, polar bears, raccoons, foxes, most birds |
| **DECOMPOSERS** | Mushrooms, bacteria |

**An Ocean Environment** The ocean is the largest environment on Earth. It also has the largest variety of living things. The surface waters get plenty of sunlight and nutrients. So these waters are full of life. However, you would need a microscope to see much of that life. It is in the form of plankton. These tiny living things float in the water. Plankton include bacteria, algae, and even tiny animals.

This sea turtle makes a meal of soft coral.

Plankton are a major part of ocean food webs. Many plankton are producers, such as algae. Snails, fish, shrimp, and other consumers eat the plankton. Still other consumers eat these consumers. In this way, energy from the sun flows throughout the entire ecosystem.

This cup coral has captured an octopus.

## Before You Move On

1. What plants, other than trees, are in a forest?
2. How does the ocean compare with land environments?
3. **Infer** Most consumers in the ocean depend on plankton for food even if they don't eat plankton. How is this true?

Energy passes through a community by moving along food chains and food webs. Producers such as plants use energy from the sun to make their food. Consumers, including animals and humans, cannot make their own food. They get energy by eating producers or other consumers. Each kind of environment contains living and nonliving things that interact. These things interact in different ways.

**Big Idea** Living things depend on each other and their environment for food.

## Vocabulary Review

**Match** each term with its correct definition.

**A. predator**
**B. prey**
**C. herbivore**
**D. carnivore**
**E. omnivore**

1. An animal that eats plants to survive
2. An animal that eats both plants and other animals to survive
3. An animal that hunts other animals for food
4. Animals that other animals hunt for food
5. An animal that eats other animals to survive

## Big Idea Review

1. **Recall** Where do most plants get the energy they need to live and grow?
2. **Define** What is a food web?
3. **Contrast** How are producers and consumers different?
4. **Explain** Why do prey need predators?
5. **Apply** What would be a common food chain in the environment where you live?
6. **Predict** Suppose that a disease of plants kills all the grasses on a prairie. How might the prairie's food web change?

## Write About Food Chains

**Explain** Choose a food chain in this food web. Describe what is happening in the food chain. Explain what is moving from one living thing to another.

CHAPTER 3

# LIFE SCIENCE EXPERT: BEHAVIORAL ECOLOGIST

## Curious about animals? How about being a behavioral ecologist?

Some people can spend hours watching how animals act. If you're one of them, you might want to be a behavioral ecologist. You could work in some of the wildest places on Earth! That's what behavioral ecologist K. Yoganand does. He talks about tracking wild sloth bears in central India.

### What does a behavioral ecologist do?

I study the behavior of wild animals in their natural environment. My goal is to find out how an animal's behavior helps it survive. Why are sloth bears most active at night? Is it because they need to avoid the daytime heat? Or is it because people are in their habitat during the daytime? The more we know about animals in the wild, the better we can protect them.

Yoganand holds a sloth bear cub while examining its health.

Yoganand (left) and an assistant put a radio collar on a sloth bear.

TECHTREK
myNGconnect.com

Student eEdition

Digital Library

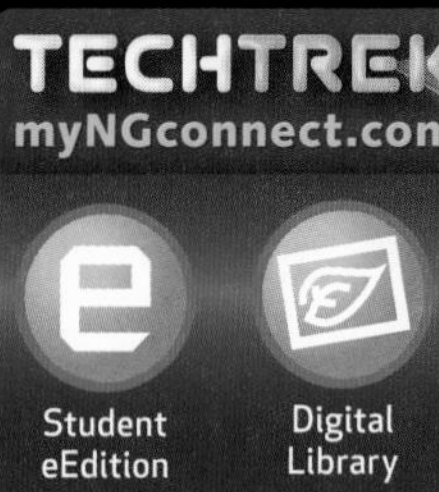

### What do you like most about your job?

I love being outdoors. Some people pay a lot of money to observe animals in an Asian forest or African savanna. It's a vacation for them. I do it as part of my job. How cool is that?

### How did you study wild sloth bears?

I spent a lot of time in the forest observing and collecting data. I wanted to find out what a typical day is like for sloth bears. When do they sleep? What do they eat? Where do they give birth? What do they do when they meet a tiger? To answer these and other questions, I had to observe them in the wild. So I put radio collars on the bears to track them.

### Was your research successful?

Yes! We now have a better idea of what people need to do to help sloth bears survive. I hope that my work helps us find ways to protect them and the land they need to live.

How do you think a sloth bear uses its long, sharp claws?

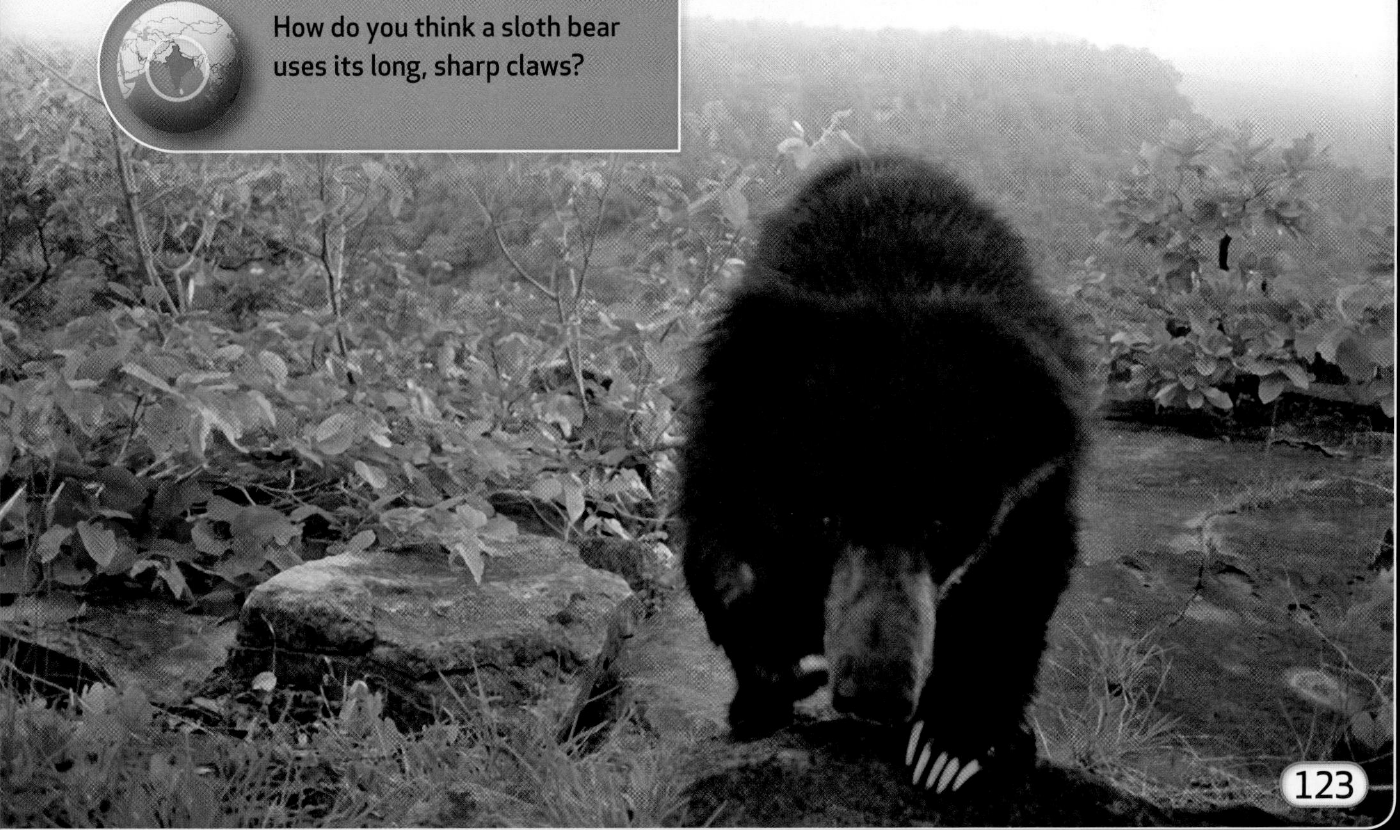

# Bats: Winged Wonders of the Night

Have you ever seen a bat flying around at night? The United States has 47 kinds of bats. But one small island in the Panama Canal has more than 70 kinds of bats. The island is Barro Colorado. How did one small place come to have so many different kinds of bats?

Small disk-winged bats use suction cups at the base of their thumb and foot to climb inside a curled leaf. These bats eat insects.

TECHTREK
myNGconnect.com
Student eEdition
Digital Library

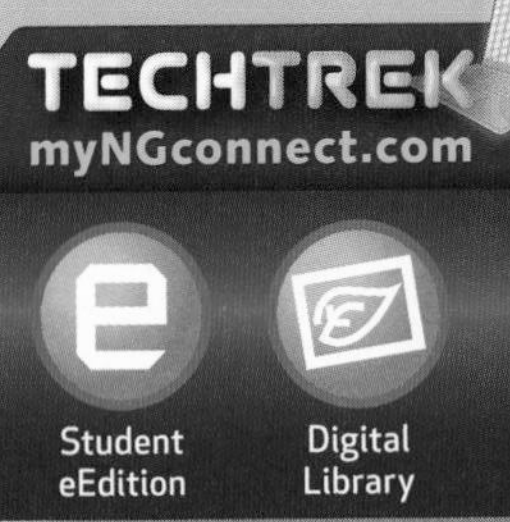

Each kind of bat became an expert in finding food a different way. By dividing the resources where they live, the bats of Barro Colorado Island become parts of different food chains and food webs. The bats avoid competing with each other for food. The following pages show some of the many different kinds of bats on Barro Colorado and around the world. Think about what each kind eats and how it finds its food.

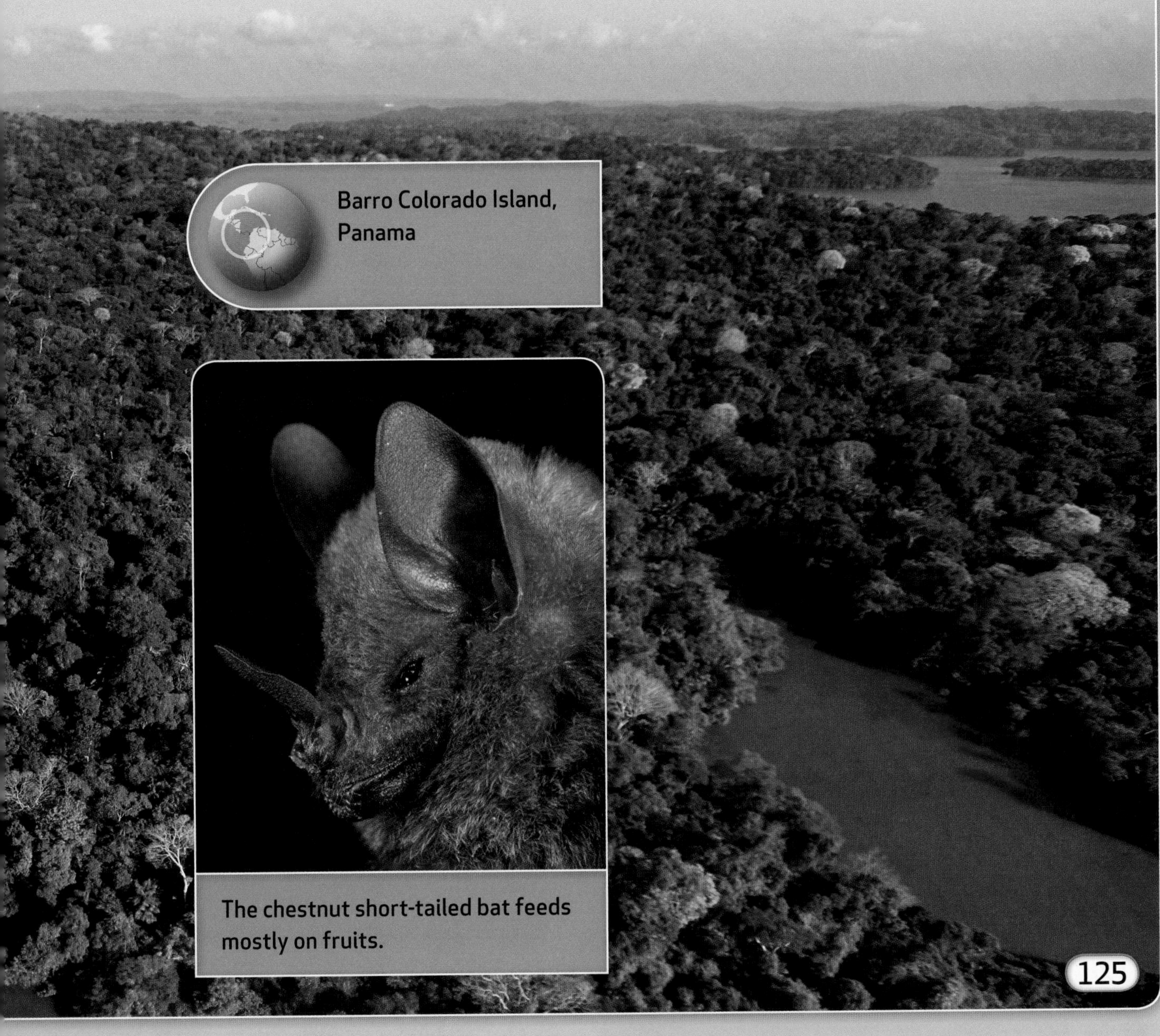

Barro Colorado Island, Panama

The chestnut short-tailed bat feeds mostly on fruits.

Like all animals, bats cannot make their own food. They are consumers. They have to eat plants or other animals to get the energy stored in that food. Bats have some special features that help them get food. These features include wings. The bones in a bat's wing are like the bones you have in your arm and hand. So a bat can use its wings not only to fly but to grasp things.

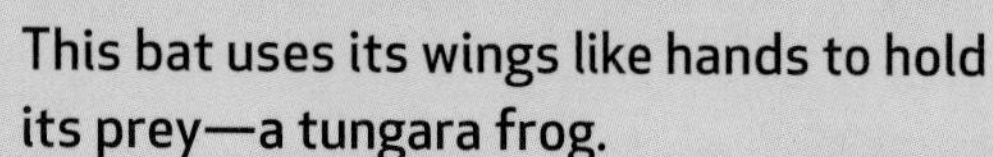

This bat uses its wings like hands to hold its prey—a tungara frog.

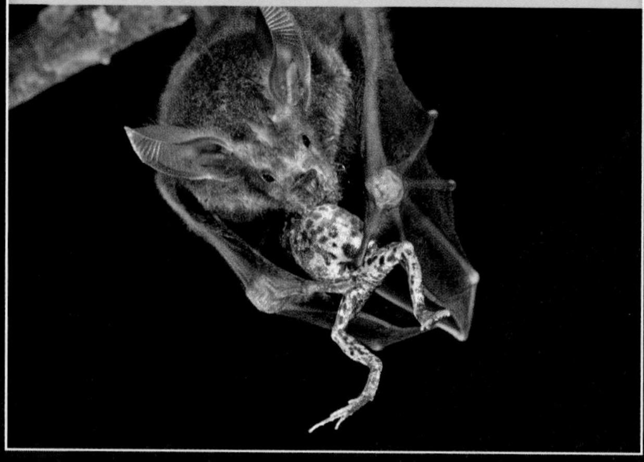

A special camera catches this bat along its flight path as it tries to snag a fish.

## Echolocation

Most bats eat at night. These bats have an extra sense called echolocation. Sound waves bounce off objects and tell the bats where the object is, how big it is, and even what shape it is. You might say the bats use sound to see. Many bats that use echolocation are predators. They hunt prey for food. But echolocation also helps bats avoid running into each other and objects such as trees.

A scientist holds a small Formosan long-eared bat.

**predator**

A **predator** is an animal that hunts other animals for food.

**prey**

**Prey** are animals that other animals hunt for food.

## Bats as Carnivores

Most bats are **carnivores**. They get energy by eating insects, fish, frogs, and other small animals. The bats use echolocation to find these animals, even fish under water. Other features, like those shown here, make the bat a successful predator.

## Vampire Bats

A few bats, called vampire bats, eat the blood of large animals, such as horses and cows. Unlike the vampires in stories, the bats do not suck blood. A chemical in their saliva keeps the blood from clotting, or hardening. Then the bat simply laps up the blood.

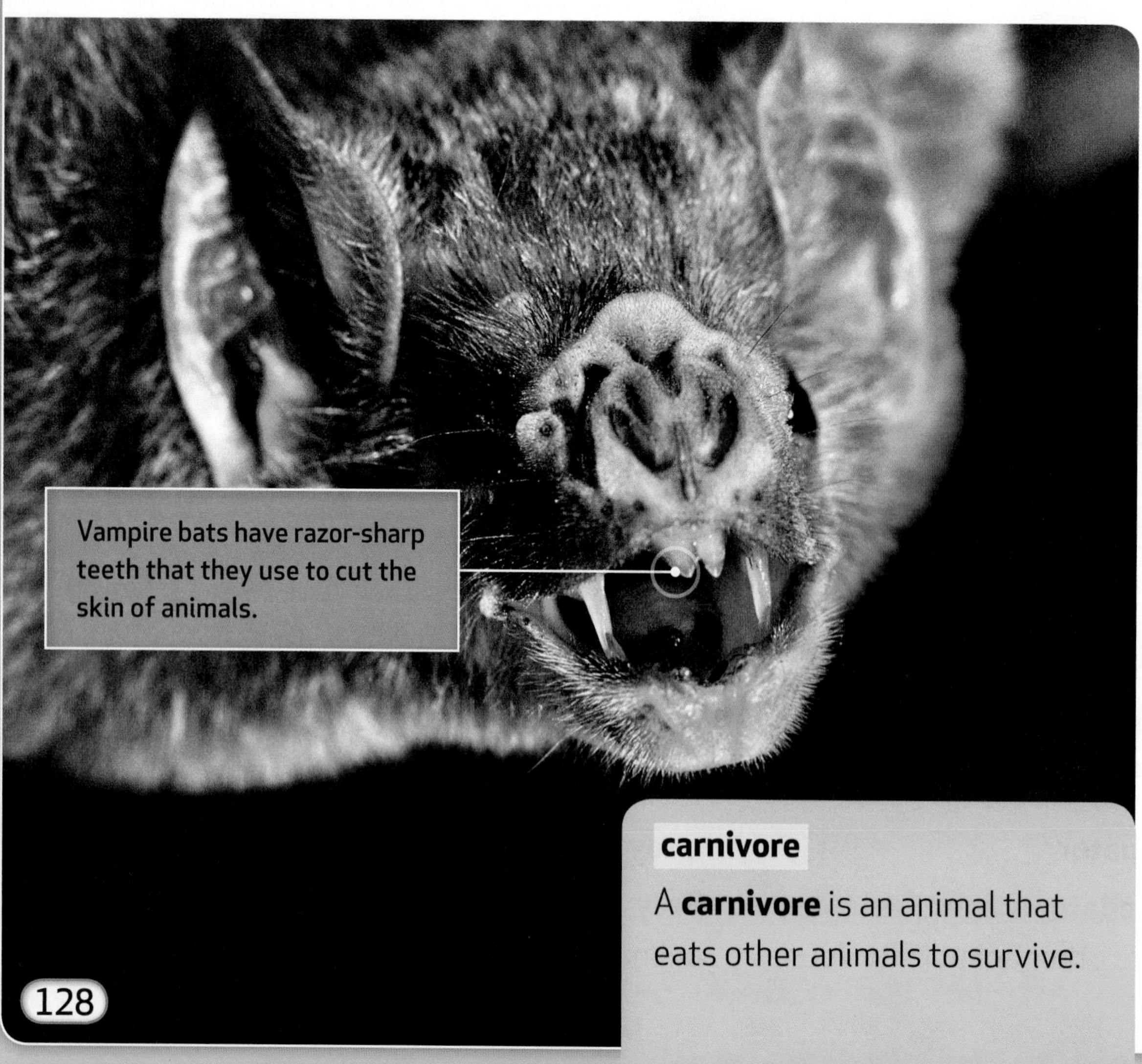


**carnivore**

A **carnivore** is an animal that eats other animals to survive.

## Bats as Omnivores

A few bats are omnivores. They eat plant parts, such as fruit, as well as small animals.

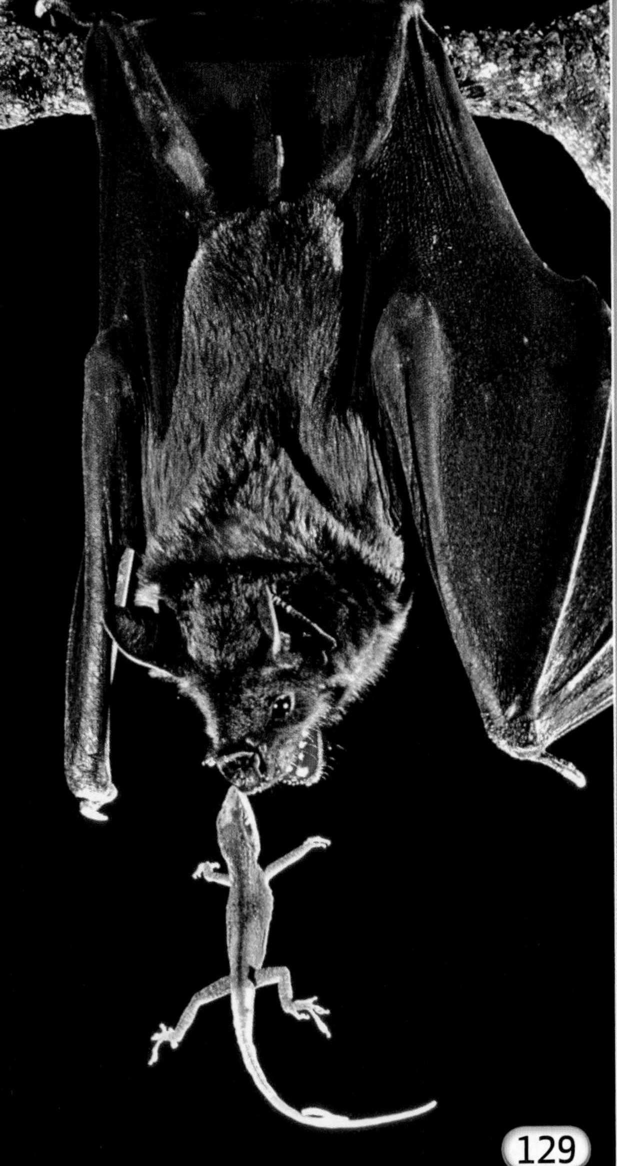

The greater spear-nosed bat is an omnivore. Here it feeds on a lizard, but it also eats fruits, insects, and pollen.

Claws on the feet of the greater bulldog bat are perfect for grabbing and holding prey.

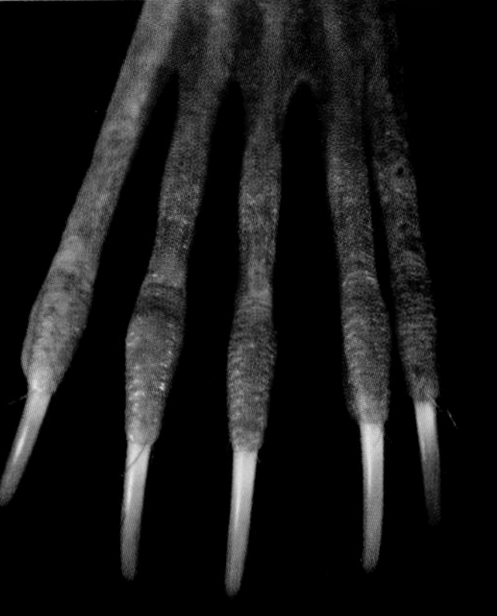

**omnivore**

An **omnivore** is an animal that eats both plants and other animals to survive.

## Bats as Herbivores

Some bats do not eat other animals at all. They get energy by eating only foods produced by plants, such as fruits and pollen grains. These bats are **herbivores**. They carry pollen from one flower to another. That allows the flowers to make seeds that grow into new plants.

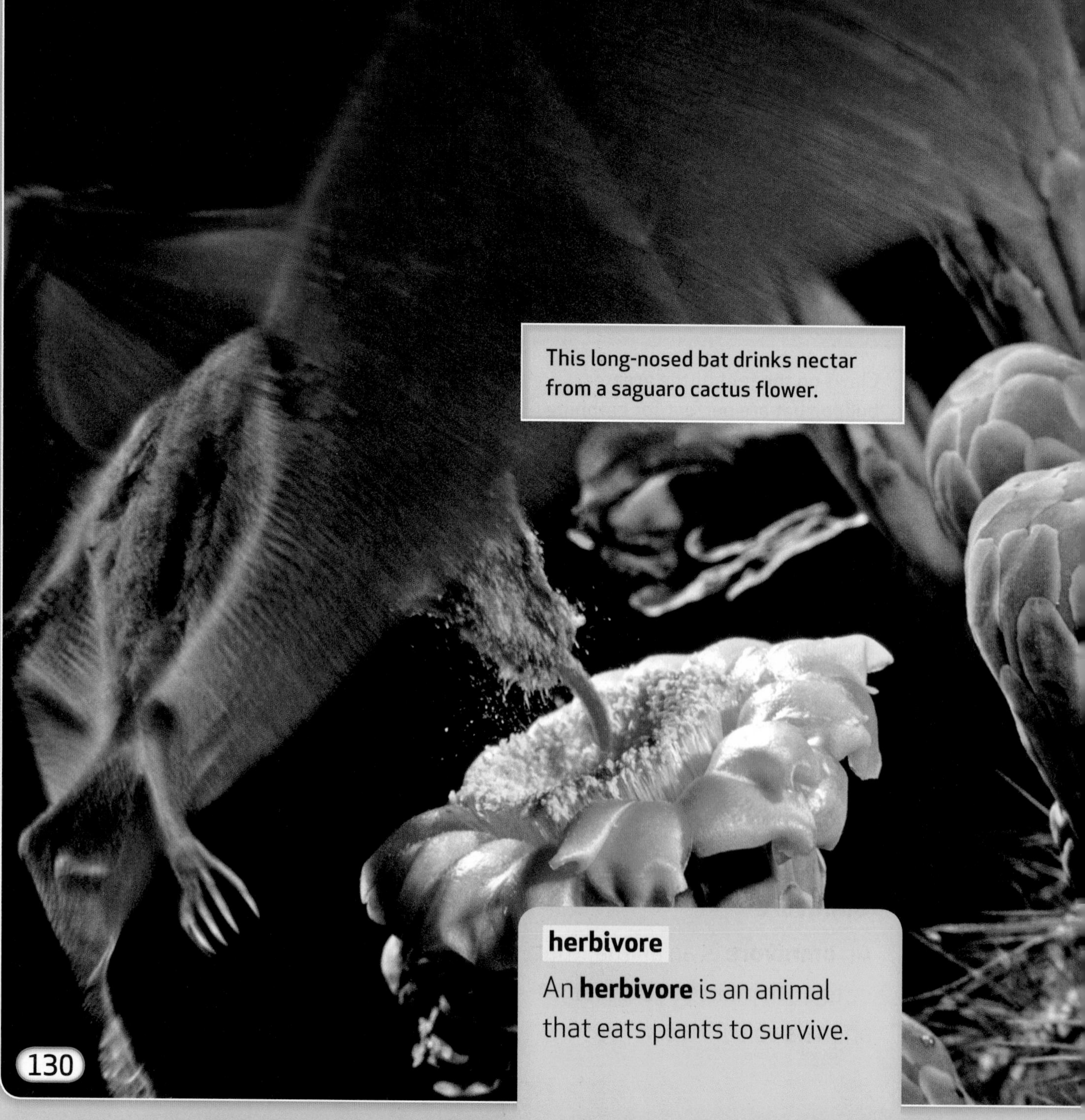

This long-nosed bat drinks nectar from a saguaro cactus flower.

**herbivore**

An **herbivore** is an animal that eats plants to survive.

The fruits of many tropical plants are food for fruit bats. These herbivores are important because they spread the seeds of the fruits they eat. Many tropical fruit crops depend on fruit bats. So do the people who eat those fruits.

TECHTREK
myNGconnect.com
Digital Library

Some people may not like the look of bats. But these winged wonders of the night are important parts of the food chains and food webs that make up a community.

NATIONAL GEOGRAPHIC

# CHAPTER 3 SHARE AND COMPARE

**Turn and Talk** Most bats feed in the dark of night. How are they able to find food or avoid running into trees? Form a complete answer to this question together with a partner.

**Read** Select two pages in this section. Practice reading the pages. Then read them aloud to a partner. Talk about why the pages are interesting.

my SCIENCE notebook **Write** Write a conclusion that summarizes what you have learned about bats. State what you think is the Big Idea of this section. Share what you wrote with a classmate. Compare your conclusions. Did you recall that different bats eat different things?

my SCIENCE notebook **Draw** Draw a picture of a bat that shows how it depends on its environment for food. Combine your drawing with those of your classmates to show a place where many different kinds of bats live.

## After reading Chapter 4, you will be able to:

- Recognize how adaptations are important to the survival of living things. **ADAPTATIONS, MOVEMENT, GETTING FOOD, DEFENSE AND PROTECTION, SENSING AND COMMUNICATING, PLANT ADAPTATIONS**
- Identify examples of animal adaptations that are used for movement, protection, sensing, communication, reproduction, and meeting the animal's needs in its environment. **ADAPTATIONS, MOVEMENT, GETTING FOOD, DEFENSE AND PROTECTION, SENSING AND COMMUNICATING**
- Identify examples of plant adaptations that are used for protection, reproduction, and meeting the plant's needs in its environment. **PLANT ADAPTATIONS**
- Recognize that modern animals and plants often resemble fossilized species. **FOSSILS AND EXTINCTION**
- Science in a Snap! Identify examples of animal adaptations that are used for movement, protection, sensing, communication, reproduction, and meeting the animal's needs in its environment. **GETTING FOOD**

CHAPTER

4

# HOW DO ADAPTAT HELP

**One secret to survival in the wild** is to avoid being seen. That's what this peacock butterfly is doing. A hungry bird might never see it. Other animals may be frightened of its wings. How does the peacock butterfly's body covering help it stay alive?

The peacock butterfly is hard to see among the colors of the autumn leaves.

TECHTREK
myNGconnect.com

Student eEdition

Vocabulary Games

Digital Library

Enrichment Activities

# SCIENCE VOCABULARY

## adaptation
(a-dap-TĀ-shun)

An **adaptation** is a feature that helps a living thing survive in its environment. (p. 138)

The flipper of this sea turtle is an adaptation for swimming in the ocean.

## camouflage (CAM-uh-flahj)

**Camouflage** is a color or shape that makes a living thing hard to see. (p. 146)

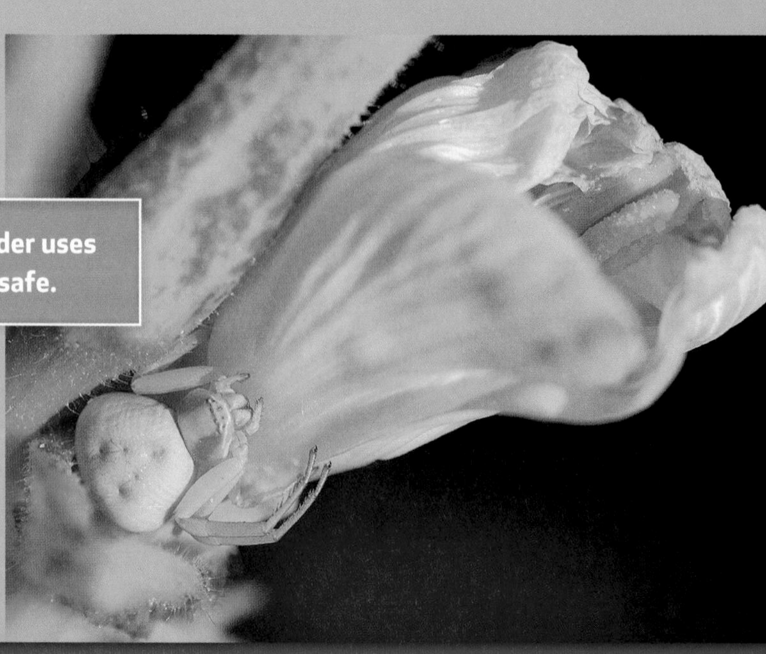

This yellow crab spider uses camouflage to stay safe.

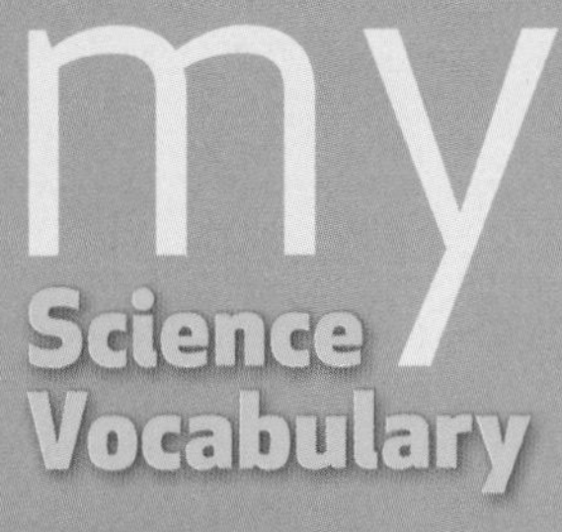


**adaptation**
(a-dap-TĀ-shun)

**camouflage**
(CAM-uh-flahj)

**extinction**
(eck-STINK-shun)

**variation**
(vair-ē-Ā-shun)

## variation (vair-ē-Ā-shun)

A **variation** is a different form of a feature of the same kind of living thing. (p. 149)

These leaf-litter toads have a color variation.

## extinction (eck-STINK-shun)

**Extinction** is the complete loss of one kind of living thing. (p. 166)

The extinction of the mastodon happened thousands of years ago.

# Adaptations

Living things come in many sizes and shapes, and their appearance is often quite unique. Look at the turtles in the photos. These animals have scaly skin, a shell, and four limbs. They are alike in more ways than they are different. Now observe the animals more carefully. The sea turtle has flippers and the tortoise has legs with feet. These limbs are features that help the animals move. A feature that enables a living thing to survive in its environment is an **adaptation**.

A sea turtle uses its flippers to swim gracefully in the ocean.

Tortoises can walk to find food. The tortoise's legs help it walk in mud or through a shallow pond. A sea turtle, though, has flippers. The flippers enable it to swim in the ocean. By swimming, the sea turtle is able to catch food and escape predators.

This tortoise uses its legs and feet to move in its environment.

## Before You Move On

1. Which part of the tortoise's body is an adaptation for walking?
2. How do sea turtles use their flippers to live in the ocean?
3. **Infer** How does a tortoise's hard shell help the tortoise survive in its environment?

ook at this kinkajou. It has several adaptations for movement.
n the ground, it walks or runs on four legs. In trees, the kinkajou raps its tail around branches to help it move. The kinkajou's il also helps it hold on while reaching for something to eat. The nkajou's tail and legs are adaptations that help it get what it eeds from its environment.

Sharp claws help the kinkajou cling to branches as it climbs a tree.

Animals have many different body parts that help them get what they need. Some body parts are adaptations for movement. Look at the photos. What adaptations help each of these animals move in the air?

## ANIMAL BODY PARTS

**This magpie uses its wings to fly.**

**Bats and some insects are other animals with wings.**

### Before You Move On

1. Name an animal that has a special adaptation for movement, and tell how it moves.
2. How are flippers and wings alike?
3. **Apply** Why can the kinkajou's tail be thought of as an extra hand?

# Getting Food

Hunger is a signal that tells you it's time to eat. Usually, the next step is to find something to satisfy the hunger. For you, that might mean a trip to the pantry for some peanut butter. But for an animal in the wild, it often means hunting for its food, or prey.

A hungry eagle uses powerful wings to hunt for food. Strong feet with claws help catch and hold its prey. Once the animal is caught, the eagle uses its beak to tear the meat and eat it.

The talons, or claws, of an eagle are an adaptation for grasping prey.

The eagle spreads its talons to catch a fish.

**Gathering and Eating Food** Most animals take in food with help from specialized body parts in their mouths. Each animal's mouth is an adaptation for the kind of food the animal eats. The giraffe has a unique way of gathering and eating its food. The giraffe uses its long neck and long tongue to grab and pull leaves off of high trees. This adaptation helps the giraffe get what it needs from its environment.

Other animals have different adaptations in their mouths that help them eat. Look at the chart below. What adaptations help each of these animals eat?

Digital Library

**A beetle slices its food with razor-like mandibles.**

**mandible**

**A chameleon has a very long tongue for getting food into its mouth.**

**Carnivores, such as this tiger, tear meat with long, pointed teeth.**

**Herbivores, such as this zebra, grind plants with large, flattened teeth.**

A bird uses its bill or beak to get food. A bird's bill is very strong. The size and shape of the bill is an adaptation for the food the bird eats.

For example, flamingos wade in shallow water. Their long bills are curved. This shape is an adaptation for scooping small animals out of the water to eat.

This duck's bill is shaped to collect food from the water.

This flamingo uses its beak to eat tiny animals called krill.

# BEAK ADAPTATIONS

Digital Library

This kingfisher's long, pointy beak helps the bird catch fish.

Insect-eaters, such as this swallow, have a thin, short beak for pinching.

The hooked beak of this bald eagle helps the bird tear meat from its prey.

Seed-eaters, such as this hawfinch, have a thick, cone-shaped beak that works like a nutcracker.

## Science in a Snap! Pick It Up!

Place objects of different sizes and shapes on a table or desk.

Try picking up the different objects with different tools, such as tweezers, pliers, or a spoon. Observe carefully which objects were easiest to pick up and why.

**Which bird shown here has a beak or bill most like a spoon? Which bird shown here has a beak or bill most like pliers?**

**Attracting and Catching Prey** Anglerfish live deep in the ocean where there is no sunlight. These fish can make their own light to attract prey. When a curious fish gets too close, it becomes food.

In the smaller picture, the flounder is waiting for small fish. The flounder uses camouflage to hide from its prey. Camouflage is a color or shape that makes a living thing hard to see. Because of the camouflage, the small fish does not see the flounder. In a flash, the small fish becomes food.

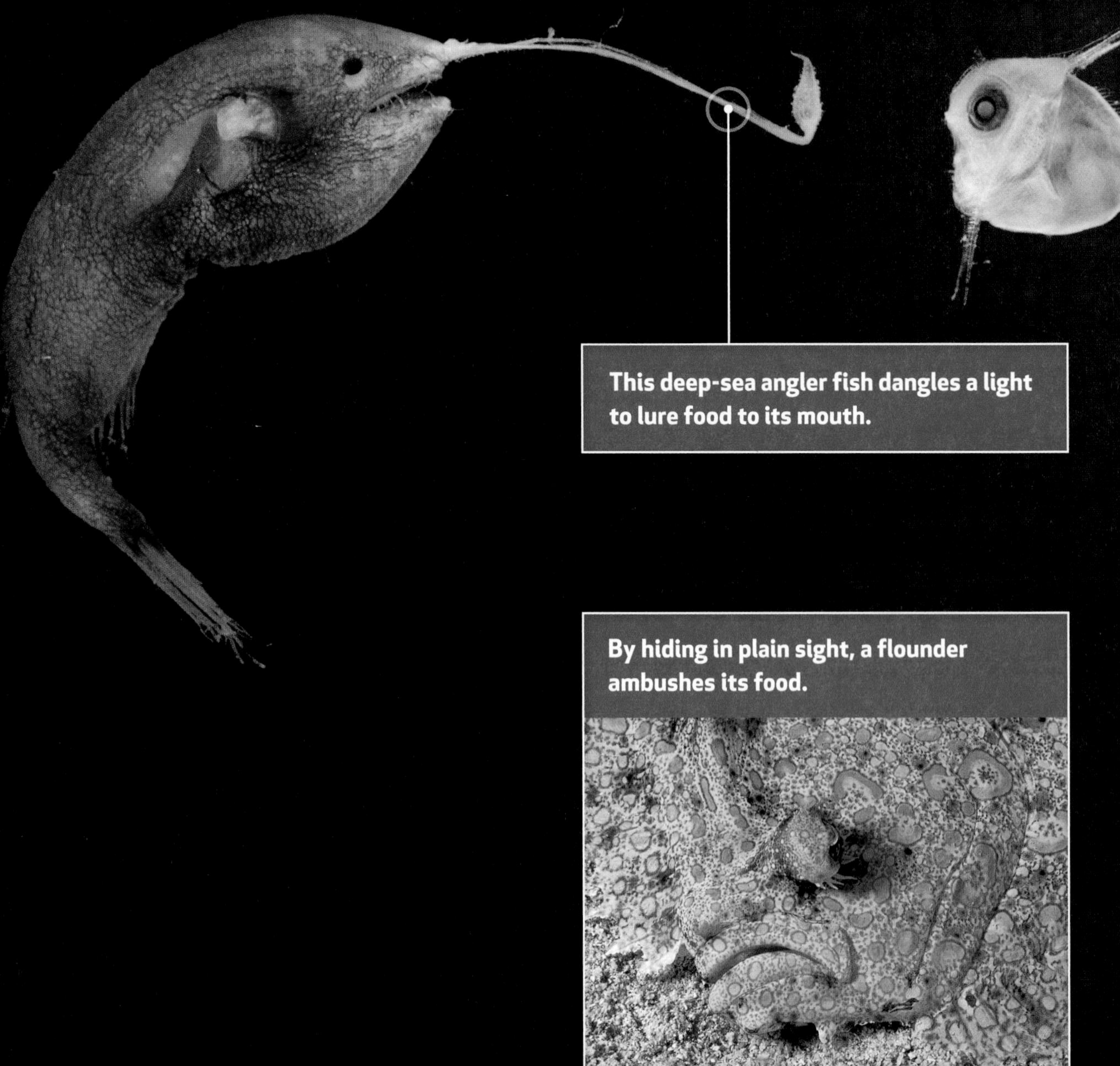

This deep-sea angler fish dangles a light to lure food to its mouth.

By hiding in plain sight, a flounder ambushes its food.

You probably know that spiders make webs to catch prey. Spinning a web is an important adaptation. But many spiders have other ways to catch food. A crab spider uses camouflage. It blends into the color of a flower. When insects land on the flower, the spider pounces on them.

A yellow crab spider blends in with the color of this flower.

## Before You Move On

1. What are three kinds of adaptations animals have for getting food?
2. How are the parts of the mouths of insects and birds alike? How are they different?
3. **Apply** A chameleon can change its body color to blend in with its environment. How does this adaptation help it catch its prey? Explain.

# Defense and Protection

What is happening between the lion and the porcupine in the picture below? The lion is probably trying to kill the porcupine. The porcupine naturally defends itself. To do so, it raises its stiff, sharp quills. If the lion gets quills stuck in its face and paws, the porcupine will get away. The quills are an adaptation to defend the porcupine against predators.

Many animals use body coverings to protect themselves. A fish's scales and a snail's shell are body coverings that protect against predators and other harm. What other body coverings protect animals?

**The lion does not attack the porcupine because of its sharp quills.**

**The African crested porcupine uses long, needle-sharp quills to defend itself.**

Find the leaf-litter toads among the leaves in the picture below. These toads are all the same kind, but they are not the same color or size. The leaves around the toads are not the same color or size, either. A **variation** is a difference in a feature of one kind of living thing. The variation of the toads helps them survive on the forest floor.

TECHTREK
myNGconnect.com

Using camouflage, leaf-litter toads hide on dead leaves that match their color variations.

Enrichment Activities

## Before You Move On

1. Besides color, what other variation do leaf-litter toads use to survive?
2. How does the porcupine use body coverings and behavior to survive?
3. **Predict** Suppose all of the leaf-litter toads were the same color. Do you think they would survive as well? Give reasons for your answer.

# Sensing and Communicating

How do you sense what's around you? Eyes for seeing, ears for hearing, and a nose for smelling are some of the body parts you use to gather information. Animals use senses to find food, avoid danger, and communicate.

The slender loris looks out into the dark night with huge eyes. Larger eyes take in more light for hunting prey at night. The loris also uses its ears to hear the sounds of any prey that might be in the area.

**This loris uses sight and hearing to hunt for insects, lizards, and birds at night.**

A great horned owl has large eyes and keen eyesight. But this owl can also use its hearing to find prey in total darkness. You might think its "horns" look like ears. But its ears are on either side of its face, near its eyes. Smooth feathers sweep out from the center of its face. They funnel sounds to the owl's ears, making its hearing even sharper.

**The great horned owl has a keen sense of hearing that helps the owl locate its prey.**

**This chameleon's eyes help it search for prey all around its body.**

**Touch and Smell** Touch and smell also help animals find what they need to survive. This bearcat smells with its nose to find food at night. Its whiskers act as organs of touch.

Many animals have antennae. Antennae are organs of touch for animals such as crayfish, crabs, and insects.

The two long stalks on this crayfish's head are antennae.

The bearcat uses its senses to find food and stay out of danger.

The star-nosed mole is an animal with a multipurpose sense organ. A ring of tentacles form a star at the tip of its nose. Some of the tentacles keep soil out of its nose as the mole digs tunnels underground. The mole uses the shortest tentacles to identify prey. The longer tentacles are the most sensitive.

A star-nosed mole finds its way with its nose instead of its eyes.

**Communicating** Step outside on a spring day and you will likely hear birds singing. They are not really "singing." They are communicating, or sharing information. Young birds make sounds to express hunger and compete for food. Adult birds make sounds to attract a mate or to tell other birds to stay away from their nest. Many animals make sounds to warn of danger, and to mark territories.

**These waved albatross use sound to communicate.**

Animals communicate using odors too. Animals have a very good sense of smell. A skunk gives off a foul odor to warn predators to stay away. Skunks and other animals also give off odors to attract mates and mark territories.

Animals also communicate by moving body parts. A skunk raises its tail to show that it is going to spray. A raised tail of a white-tailed deer warns other deer that danger may be near.

**When a skunk raises its tail, it is about to release a very strong odor.**

**Reproduction** This peacock is doing his best to attract a peahen. His strutting back and forth and his colorful tail feathers are adaptations. The color of the peacock's tail is an adaptation that helps it survive and reproduce. When living things reproduce, they pass along many of the adaptations that help the offspring survive.

A peacock displays his tail feathers to attract a mate.

Two male bighorn sheep fight to decide which is stronger. This type of contest is an adaptation for reproduction in many animal species.

In some places, fireflies glowing at night are a common sight in summer. These insects give off light to attract mates. Many animals that live deep in the ocean, such as small shrimp and marine worms, also have this adaptation.

Male and female fireflies communicate by flashing signals to each other.

## Before You Move On

1. What might birds be communicating when they "sing"?
2. Sound is one way that animals communicate. What are other ways?
3. **Infer** What might you think a dog is communicating when it growls?

# Plant Adaptations

You may think you're just looking at a patch of yellow flowers, but you are very wrong. Each of these "flowers" is a specialized leaf for trapping insects. Pitcher plants and Venus flytraps capture and digest small animals such as insects and absorb nutrients from them. As you can see, plants have adaptations for survival, too.

The leaves of the pitcher plant form a tube. Insects fall into the tube and cannot get out.

Pitcher plants grow in soil that lacks nitrogen. The plants get this nutrient from animals.

When an insect lands on a Venus flytrap, it touches tiny hairs that trigger the leaves to snap shut. The plant absorbs nitrogen as the insect decays.

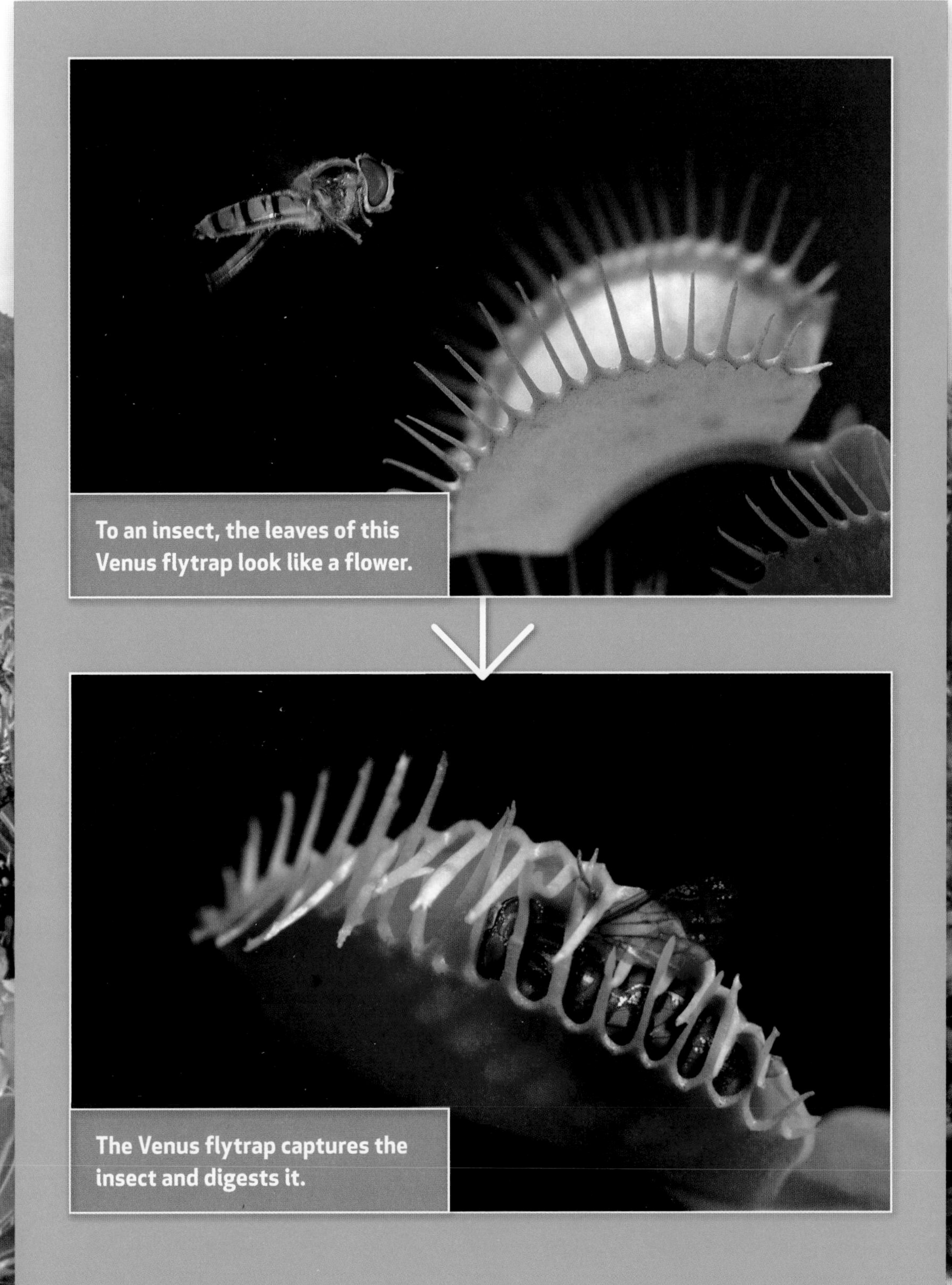

To an insect, the leaves of this Venus flytrap look like a flower.

The Venus flytrap captures the insect and digests it.

**Needles and Thorns** You might be surprised to learn that the needles of pines and firs and the spines of a cactus are leaves. These kinds of leaves are adaptations for reducing water loss. The small surface area of each needle reduces the amount of water that evaporates from inside the needle.

The colorful plant shown here grows in tropical climates. Look closely. The plant is armed with thorns. These structures are adaptations to protect the plant against animals that might eat the plant.

The thorns of a bougainvillea help protect the plant.

**Leaves and Stems** A jade plant lives in a hot, dry climate. The plant's thick leaves and stems store water. They have a waxy covering that prevents water loss. The jade plant can survive long periods without water.

Some plants live in wet places. These plants often have large, thin leaves, which help them lose water through evaporation. The size and shape of the leaves are adaptations for surviving in the environment.

**The jade plant's leaves and stems help it survive in very hot, sunny places.**

**Reproduction** The flowers of a colorful bird-of-paradise plant attract hummingbirds looking for food. The long, tubelike bill of the hummingbird is adapted for reaching the flower's nectar. While it eats, the bird picks up some of the flower's pollen. When the bird flies to another flower of the same kind, the pollen rubs off the bird onto that flower. This pollination starts a series of steps that makes seeds and new plants. In this way, flowers are adaptations for reproduction.

The bird-of-paradise flower and the hummingbird's bill are adaptations for pollination.

After pollination, plants produce seeds. A seed contains an offspring of a plant. Most seeds develop inside of a fruit. A seed's best chance to grow into a healthy plant is to be spread, or dispersed, away from the parent plant. Fruits have structures that help do that. When seeds disperse, young plants will not have to compete with their parents for resources such as water, light, and nutrients.

## **ADAPTATIONS** FOR SEED DISPERSAL

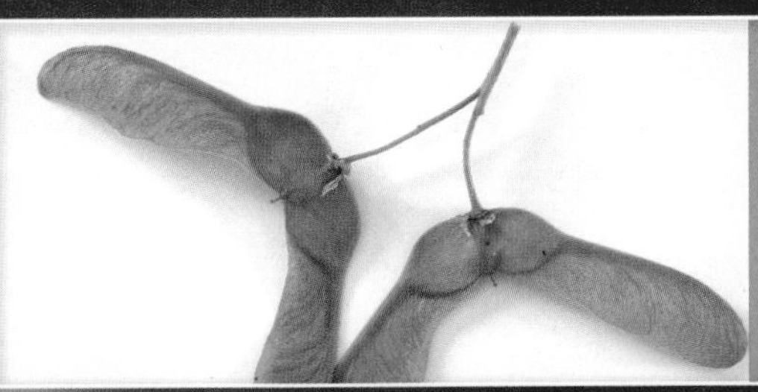

### BY WIND

**These wings carry away these four maple seeds on the wind.**

### BY ANIMALS

**The fruit of a burdock is covered with hooks that cause them to stick to animal fur.**

### BY WATER

**A coconut palm's seeds are partly hollow and float away on water.**

### Before You Move On

1. What are three adaptations of leaves and stems that help plants survive?
2. Why is the dispersal of seeds important for the survival of plants?
3. **Predict** Could a pitcher plant survive without trapping insects in its leaves? Why or why not?

# Fossils and Extinction

If you look closely at some rocks, you might discover fossils known as trilobites. A fossil is the remains or trace of a living thing from long ago. Most fossils form in layers of rock, though some fossils form in ice and even in hardened tree sap.

Fossils help us learn about Earth's history. They show how Earth has changed over time and how it may continue to change. For example, the trilobite fossil shown here is from South Dakota. From this find and other evidence, scientists are able to tell that a shallow ocean once covered South Dakota.

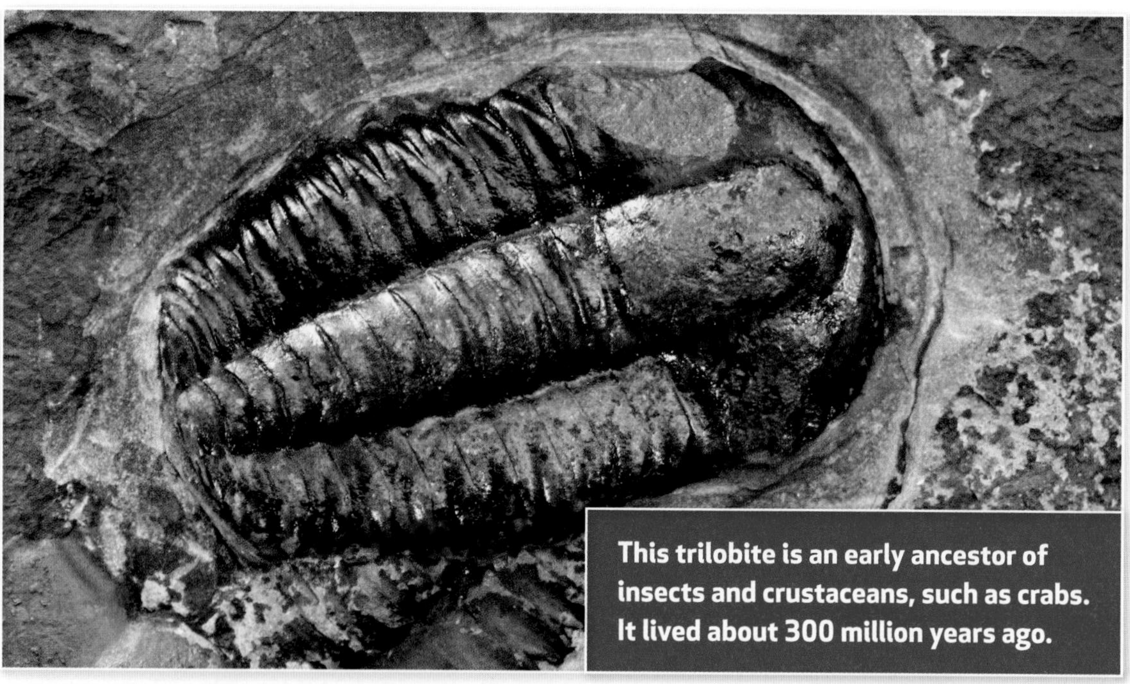

**This trilobite is an early ancestor of insects and crustaceans, such as crabs. It lived about 300 million years ago.**

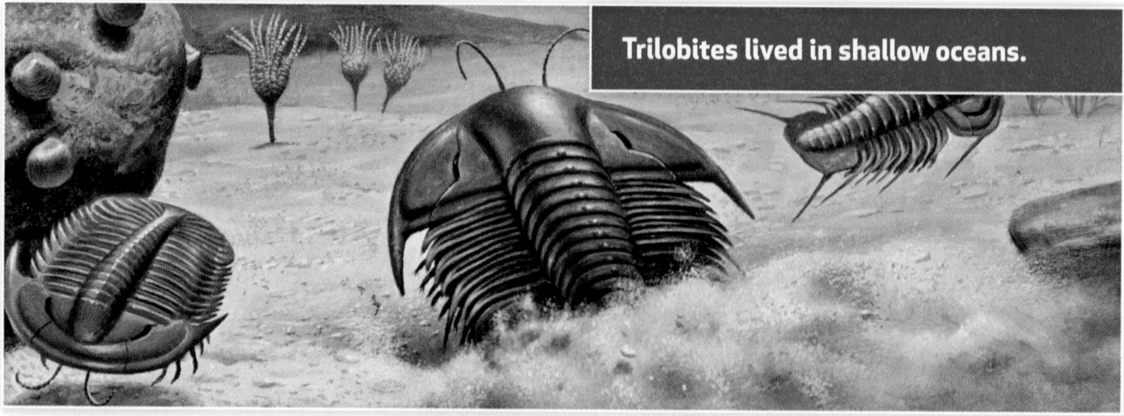

**Trilobites lived in shallow oceans.**

Do you recognize this fossil as a leaf from a fern? The fossil looks much like the leaves of modern ferns. When this plant was growing millions of years ago, ferns lived in swamps covering very large areas. Today, ferns still grow in swamps. But other ferns survive in dry places. These ferns developed adaptations that helped them survive as conditions changed.

TECHTREK myNGconnect.com

Digital Library

This fern leaf fossil has been in this rock for millions of years.

The leaf of a modern fern resembles the fossil shown above.

**Comparing Living and Extinct Organisms** You won't find trilobites swimming in the ocean today. They are extinct. **Extinction** is the complete loss of one kind of living thing. Although trilobites are extinct, many of their relatives, such as crabs, crayfish, and lobsters, remain. These animals have adapted to the present conditions of their environment, but they share some of the same traits as trilobites.

Many other extinct organisms are closely related to organisms living today. For example, if you compare the modern elephant to the American mastodon, you will see similarities. Mastodons lived during the last ice age. They became extinct over 10,000 years ago when the climate warmed.

**The elephant is a distant relative of the extinct mastodon.**

We know about mastodons because of their fossilized bones. Teams of scientists have dug up many remaining bones. These scientists then have assembled the bones back together into a skeleton. They compare the skeleton of the extinct animal with its living relatives. Comparing fossils with living organisms helps scientists learn why one kind of living thing survived and another kind did not.

**Mastodons have many of the same adaptations as the modern elephant.**

**The skeleton of this mastodon shows tusks similar to the modern elephant.**

## Before You Move On

1. How do fossils help us tell Earth's history?
2. What are some similarities and differences between mastodons and elephants?
3. **Draw Conclusions** Hundreds of different kinds of sharks live in the ocean today. Sharks have lived on Earth for about 400 million years. What does this tell us about the adaptations of sharks?

# CAMELS AND PEOPLE: SURVIVING IN THE SAHARA

The native people of the Sahara are nomads. They move from one place to another, carrying all that they have with them. Camels are a large part of the nomads' way of life. A camel is a large, grazing mammal of the desert. For more than 4,000 years, camels have carried the Saharan people and their goods across the desert. The people use the milk and meat of camels. They also weave camel hair into cloth, rope, and rugs.

A nomadic herder rides a camel that also carries the herder's belongings.

Camels have several adaptations for desert life. Their two-toed feet are wide, flat, and soft, which helps them walk on loose sand. They can go for a week without food and even longer without water. When they do drink, they can take in up to 145 liters (32 gallons). Much of this water goes into the camel's bloodstream. What about the camel's hump? It doesn't store water; it stores fat. This fat supplies energy for a long trek through the desert.

**Camels can live for many days without food or water.**

**Camels are known as "ships of the desert" for their role in transportation.**

## Conclusion

All living things have adaptations. Some adaptations help animals move. Others help them get food. Camouflage and other adaptations protect animals from predators and help them catch prey. Animals and plants also have adaptations that help them reproduce.

**Big Idea** Adaptations are features that help living things survive in their environment.

Owls have adaptations that help them survive and thrive in their environment.

## Vocabulary Review

**Match** each of the following terms with the correct definition.

**A. extinction**
**B. camouflage**
**C. variation**
**D. adaptation**

1. A color or shape that makes a living thing hard to see
2. A feature that helps a living thing survive in its environment
3. A different form of a feature of the same kind of living thing
4. The complete loss of one kind of living thing

## Big Idea Review

1. **Recall** What are two adaptations animals have for moving in water?

2. **Restate** What do scientists learn about Earth and its living things by studying fossils?

3. **Relate** What types of body coverings do some animals and plants have? How do the animals and plants use these coverings?

4. **Explain** How do animals use adaptations for sensing and communicating?

5. **Evaluate** Why are physical adaptations important to all living things?

6. **Analyze** Scientists collect seeds from all kinds of plants. They store the seeds in safe places called seed banks. What is the purpose of storing all kinds of seeds? Why would this be useful?

## Write About Adaptations

**Compare and Contrast** What adaptations do the animals in these photos display? Which ones are used for similar reasons? What are some differences in the other adaptations shown?

# Kate Jackson, Herpetologist

**What does a herpetologist do?**

A herpetologist is an expert on amphibians and reptiles, but herpetologists do many different jobs. A herpetologist could take care of reptiles in a zoo, or work outdoors learning about amphibians in a particular pond.

**What do you remember liking about science when you were in elementary school?**

When I was in elementary school, I knew that I loved animals—especially ones like snakes and toads that most other people don't like. However, I didn't realize that this fascination had anything to do with science, or a possible career studying these animals.

**When you were younger, did you ever see yourself doing what you do now?**

When I was a little girl growing up in Toronto, Canada, I thought I was the only person in the world who was passionate about snakes. Then I found out about the Ontario Herpetological Society. I learned a huge amount about herpetology from being part of that group.

TECHTREK
myNGconnect.com
Digital Library

Kate Jackson

A harmless gopher snake about to attack!

**What has been the coolest part of your job?**

I love the excitement of catching snakes in remote places where no other herpetologist has been before. I also love the excitement of finding a species that hasn't been seen in 100 years, or one that might be a new species.

**Where do you hope your work has made the biggest impact?**

I became a herpetologist because I love snakes (and other herps). Through my training as a scientist, I learned that snakes, like all living organisms, are part of ecosystems, and that other organisms (including us) depend on those ecosystems for survival. I hope that my research in Africa will add to the scientific knowledge that we need to prevent unappreciated animals from going extinct. Just as importantly, I hope that through my teaching, I can influence other people to understand the importance of, and care about, snakes and other neglected animals.

Herpetologists study all kinds of reptiles, such as these crocodiles in Singapore.

NATIONAL GEOGRAPHIC

BECOME AN EXPERT

# Snakes: Adaptations for Many Environments

Snakes are one of the most common animals on Earth. There are over 2,000 species of snakes, and snakes have been on Earth for more than 150 million years. Obviously, snakes have **adaptations** that help them survive and thrive in their many environments.

Like most reptiles, snakes have dry skin covered by scales. Unlike other reptiles, however, snakes do not have legs. Even so, scientists are sure that an ancestor of modern snakes had legs. Boa constrictors, like the one shown here, have very small hip and leg bones near the tail. The only outward signs of these body parts are two claws. The loss of legs is just one of many amazing adaptations that snakes have to help them live in their environments.

A boa's claws are connected to tiny leg bones.

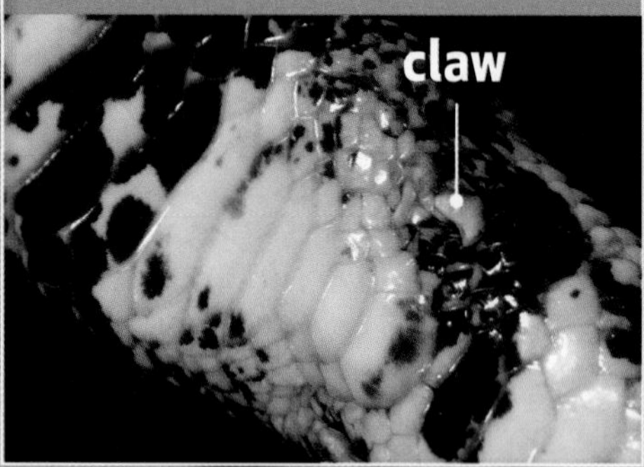

**adaptation**

An **adaptation** is a feature that helps a living thing survive in its environment.

## Snake Senses

Did you ever wonder why a snake moves its tongue in and out so often? It's smelling things. The tongue gathers scents from the environment and brings them to sense organs inside the snake's mouth.

What else is unusual about snake senses? You may know that snakes can feel vibrations on the ground, so they can feel if you're walking nearby. But can they hear you? Snakes can hear, but their ears are inside their body.

Snakes that are active in daylight have good vision. Snakes called pit vipers, however, have poor eyesight. Pits on their heads are an adaptation for sensing heat. They sense heat to "see" in the dark.

A boa constrictor moves quietly along the ground.

The fork in a snake's tongue helps the snake tell which direction a smell comes from.

## Getting Food

All snakes eat animals and swallow their food whole. Their teeth point into their mouths. This adaptation keeps prey from getting away.

But how do they catch their prey? Snakes surprise and disable their prey. Venomous snakes inject a poison that kills prey. Constrictors bite their prey and then wrap their bodies around it. Their powerful muscles squeeze the prey to kill it.

This young python is a constrictor that kills its food by squeezing it.

Snakes like this brown house snake can swallow large prey.

## Defenses

Besides rattling, a rattlesnake rises when it feels threatened, usually by a larger animal. This movement lets the other animal know that the snake is ready to defend itself. These behaviors help snakes avoid **extinction**. Snakes are more likely to survive if they avoid danger.

Different snakes defend themselves in different ways. A cobra rises up and spreads the ribs near its head. A coral snake's colors scare away possible danger. Other kinds of snakes mimic these defenses.

The hood of a cobra makes it look much larger and scarier to predators.

TECHTREK
myNGconnect.com

Digital Library

Most animals know that the colors of a coral snake mean danger.

**extinction**

**Extinction** is the complete loss of one kind of living thing.

## Camouflage

Can you see the deadly gaboon viper in the photo? It is hidden in the leaves on the ground. Many snakes have colors that help them blend in with their background. **Camouflage** helps them surprise prey and avoid predators. Many other types of snakes use camouflage too. They have many different patterns and colors. **Variations** help them to stay hidden in different environments.

The gaboon viper is one of the most deadly venomous snakes.

A smooth green snake finds grasshoppers by hiding in grass.

**camouflage**

**Camouflage** is a color or shape that makes a living thing hard to see.

**variation**

A **variation** is a different form of a feature of the same kind of living thing.

## Living in Their Environment

Snakes are adapted to the environment in which they live. A snake's adaptations help it find food and protect itself. Snakes have adapted very well to a variety of environments, and can be found on every continent on Earth except Antarctica.

These pythons are adapted to the jungles of Papua New Guinea.

# SHARE AND COMPARE

**Turn and Talk** What kinds of adaptations do snakes have that help them survive in their environment? Form a complete answer to this question together with a partner.

**Read** Select two pages in this section. Practice reading the pages. Then read them aloud to a partner. Talk about why the pages are interesting.

my SCIENCE notebook **Write** Write a conclusion that summarizes what you have learned about the adaptations of snakes. In your conclusion, restate what you think is the Big Idea of this section. Share what you wrote with a classmate. Compare what each of you wrote. Did you recall examples of different snake adaptations?

my SCIENCE notebook **Draw** Snakes defend themselves in many different ways. Form groups of three. Each person in the group should draw a picture of a snake defending itself in a different way. Compare what each of you drew. Post your drawings in the classroom.

## After reading Chapter 5, you will be able to:

- Give examples of how plants change based on the seasons. **CHANGES IN SEASONS**
- Explain how plants and animals adapt to the changing of the seasons in their environment. **CHANGES IN SEASONS**
- Give examples of how animals adapt to the changing of the seasons in their environment. **CHANGES IN SEASONS**
- Explain how plants can change the environment in many ways. **PLANTS CHANGE THE ENVIRONMENT**
- Describe how plants can harm an environment. **PLANTS CHANGE THE ENVIRONMENT**
- Explain how animals can change the environment by making homes, changing soil, grazing, and feeding. **ANIMALS CHANGE THE ENVIRONMENT**
- Describe how animals can hurt the environment. **ANIMALS CHANGE THE ENVIRONMENT**
- Observe and explain that humans can change the environment in beneficial or harmful ways. **PEOPLE CHANGE THE ENVIRONMENT**
- Recognize and explain that humans depend on their environments to meet their needs. **PEOPLE CHANGE THE ENVIRONMENT**
- Science in a Snap! Observe and explain that humans can change the environment in beneficial or harmful ways. **PEOPLE CHANGE THE ENVIRONMENT**

CHAPTER

5

# HOW DO LIVING THINGS INTERACT ENVIRON

**Everything on Earth** is connected, like the threads of a web. Change one thing, and you affect something else. For example, as winter changes to spring, these birds gather on high cliffs to mate and raise their young. How do the birds interact with their environment to live? How do these interactions change the environment?

These seabirds are northern gannets. Thousands of them gather on cliffs on Bonaventure Island, Quebec in Canada each spring.

TECHTREK
myNGconnect.com
Student eEdition
Vocabulary Games
Digital Library
Enrichment Activities
WITH THEIR
MENT?

# SCIENCE VOCABULARY

## invasive organism
(in-VĀ-siv OR-guh-niz-uhm)

An **invasive organism** is a plant or animal that does not belong in a place and harms it. (p. 194)

Kudzu is an invasive organism that can take over a forest.

## conserve (kun-SURV)

To **conserve** is to use resources in a careful way. (p. 205)

Using recycled materials to build homes is one way to conserve natural resources.

**conserve**
(kun-SURV)

**invasive organism**
(in-VĀ-siv OR-guh-niz-uhm)

**pollution**
(pah-LU-shun)

**recycle**
(rē-SĪ-kuhl)

## recycle (rē-SĪ-kuhl)

To **recycle** is to use something again. (p. 205)

One way to help conserve natural resources is to recycle cans.

## pollution (pah-LU-shun)

**Pollution** is a harmful substance that people put into the air, water, and soil. (p. 208)

Oil spills cause pollution that harms wildlife.

# Changes in Seasons

**Changes in Plants** Have you ever heard the saying, "April showers bring May flowers"? At different times of the year, you can see how changes in the environment have changed plants. During the spring, flowers bloom and leaves grow on trees. Some plants grow fruit during the summer. In the fall, leaves on some trees change color. Then the leaves drop. The trees stay bare during the winter.

**Some trees drop their leaves in the fall. New leaves will grow in the spring.**

Trees may not change this way where you live. Trees and other plants change more during the seasons in the northern part of the United States than in the southern part. However, even some southern plants show seasonal changes. Some grasses in Florida, for example, turn brown or purple in the fall. The green leaves of dogwood trees and red maples turn red and purple.

TECHTREK
myNGconnect.com
Digital Library

During the summer, green grapes grow on the sea grape plant along the Florida coast.

The grapes can turn red and orange in late summer or early fall.

**Changes in Animals** Plants are not the only living things that change with the season. Animals also must adapt to seasonal changes in order to survive. Some animals, such as deer, grow thick coats of fur to keep warm in the winter. During the summer, they shed some of their fur to stay cool. The fur of some animals changes color with the seasons. The change of color helps the animal blend in with its surroundings.

**In the spring and summer, the fur of a snowshoe hare is brown.**

**In the winter, the fur of the snowshoe hare turns white. This helps it blend in with the snow and hide from predators.**

Animals change in other ways to survive changes in seasons. During the late summer, some animals eat large amounts of food. Their bodies store the extra energy. They use that energy during the cold winter while they hibernate, or go into a deep sleep.

Other animals migrate, or move, to warmer places during the fall. Many northern birds migrate because there is not enough food in their summer homes during winter.

**Woodchucks hibernate during the winter to use less energy.**

**Woodchucks become active again in late February and March.**

The birds shown here are roseate spoonbills. Like other birds, they build a nest to get ready to lay eggs. Most birds build their nests in early spring because the weather is getting warmer. Warm weather helps young birds survive. However, spoonbills start building nests in November. This is the beginning of the dry season in Florida, where most spoonbills live. With less rain, the water level in lakes and rivers goes down. That makes it easier for the birds to gather fish and insects to feed their young.

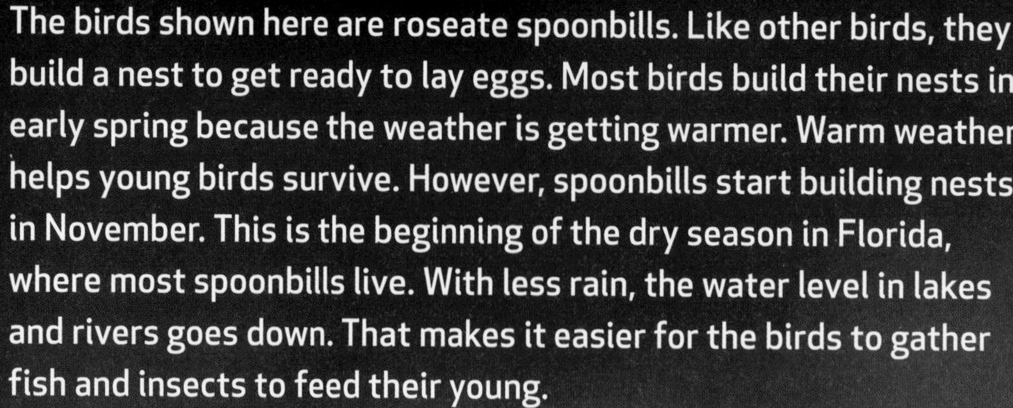

Roseate spoonbill chicks peck at food in their parent's bill.

The alligator in the photo copes with dry weather in a different way—by digging gator holes. These holes collect water from the surrounding swamps and marshes. So the holes almost always have water, even during dry weather. Alligators hunt the animals that come to drink from these pools of water.

TECHTREK
myNGconnect.com

During dry weather, the only wet places in parts of the Florida Everglades are gator holes.

Digital Library

## Before You Move On

1. What are some ways that plants are affected by the environment?
2. What are some ways that animals adapt to seasonal changes?
3. **Infer** How might snowshoe hares be affected if their fur didn't change color in winter?

# Plants Change the Environment

**Changing the Land** The changing seasons show how the environment affects plants and animals. But this cause-and-effect relationship works both ways. Plants and animals also affect the environment. Even something as simple as grass affects the world around you. The grass and other green plants give off gases that we use to breathe. The grass roots also hold the soil in place. The tiny roots form a net that keeps the soil from blowing away or washing away.

Yellow sea oats help keep this sand dune in place.

Plant roots not only keep soil in place—they help build it up. The roots trap tiny pieces of rock and decaying plant and animal matter. These materials slowly add to the soil. Roots even break apart rocks. Roots often grow into the cracks of rock. As the roots grow, they push against the sides of the cracks. The cracks widen and split the rock.

**Tree roots are splitting this boulder in Rock High Falls Gorge, New York.**

**Invasive Plants** The water hyacinth plants shown below have beautiful flowers. That's why people first brought these plants from South America to the United States in the 1880s. The plants float on water, and people thought they looked nice on ponds. But water hyacinth grows quickly. It can cover a pond or marsh and kill fish and other plants. Water hyacinth is an **invasive organism** —a living thing that doesn't belong in a place and can harm it.

Water hyacinth covers part of the Myakka River in Florida.

The purple-pink flowers of water hyacinth are beautiful, but the plant can harm the environment.

Another invasive plant is kudzu. People brought kudzu to the United States from Asia. They wanted to use it to cover bare hills and stop soil from blowing and washing away. It worked. But kudzu grows so quickly that it crowds out and covers other plants. Kudzu has spread all over the southeastern United States.

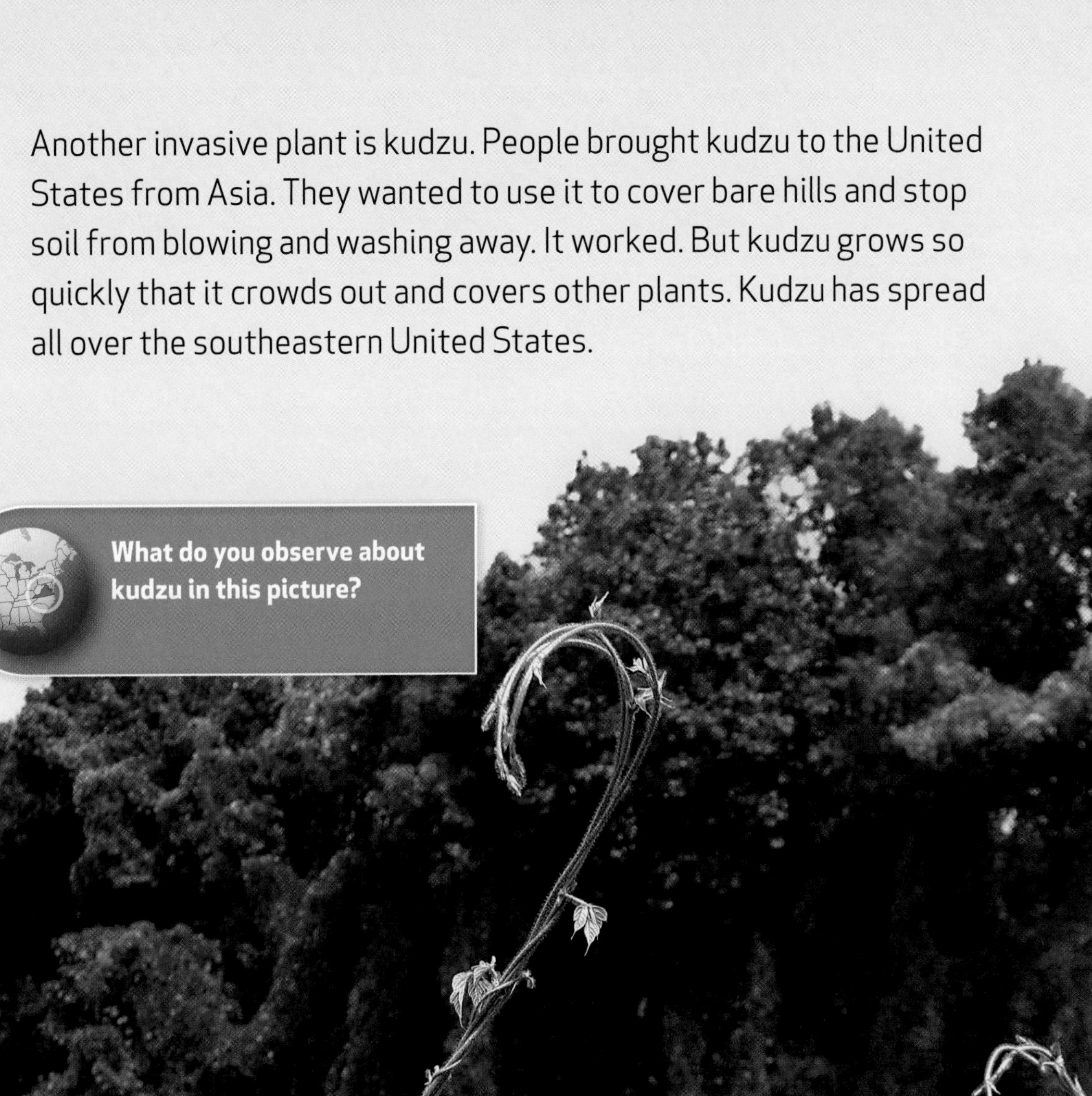

What do you observe about kudzu in this picture?

## Before You Move On

1. What is an invasive organism?
2. How do plants help build and protect soil?
3. **Generalize** How do you think kudzu harms other plants when it covers them?

# Animals Change the Environment

**Making Homes** Have you ever been to a construction site? You can see how people change the environment when they build homes. Animals change the environment when they build homes too. For example, beavers use their large teeth to cut down trees. They use the trees to build dams across streams. These dams block the flowing water. The water backs up and forms ponds and wetlands. Beavers build their lodges, or homes, in this deeper water.

TECHTREK
myNGconnect.com
Enrichment Activities

This beaver pushes a branch through the water to build its lodge.

Large amounts of branches and mud went into making this beaver lodge in Colorado.

**Changing Soil** When plants and animals die, their bodies break down, or decay, and become part of the soil. Living things called bacteria eat the decaying bodies. Then earthworms and insects eat the bacteria in the soil. When bacteria and insects die, they decay into the soil as well. All these living things help make new soil and keep it clean. When worms and insects dig through the soil, they make spaces that water can seep into. Plants growing in the soil depend on these changes.

Some termites make mounds out of soil and their own wastes. Some of these termite mounds are taller than a person.

Inside a termite mound, the tiny insects are busy building and eating.

**Feeding** Animals affect the environment just by eating. For example, cattle eat grass. Sometimes cattle graze too much in one place. They might dig up the roots or kill the grass and other plants. Plant roots help hold bits of soil together. If plants die, there is nothing to hold the soil in place. Wind and water can easily wear it away.

The picture shows another animal that changes the environment when it eats. Caterpillars eat huge amounts of leaves. They can kill trees and other plants, including crops that people grow for food.

**Caterpillars interact with the environment when they feast on tree leaves. How does this interaction change the environment?**

If tiny caterpillars can change the environment, imagine what elephants can do! Elephants eat only plants. These huge animals often tear bark and branches off trees or knock them over. But elephants help the environment too. They spread the seeds of trees when they drop their wastes in new places. Elephants use their tusks to dig for water under the ground. Other animals drink from the water holes they make.

TECHTREK
myNGconnect.com
Digital Library

These African elephants feed on bark from an acacia tree.

**Invasive Animals** You know that invasive plants can harm the environment. Invasive animals also cause harm. For example, people brought nutria from South America to the United States to raise them for their furs. Nutria have been released into the wild in many places. They eat plants that grow in wetlands. When nutria eat too many plants, no roots are left to hold soil in place. Then the soil washes away quickly.

Nutria can destroy wetlands by eating wetland plants.

The picture shows an invasive animal in Florida that is harmful in many ways—a Burmese python. This huge snake comes from the rainforests of southeast Asia. Some people keep pythons as pets. But when the snakes get too large, people sometimes release them into the wild. This is a problem. Like most invasive species, pythons have no natural enemy in their new home. So they live to reproduce and grow in large numbers. Hundreds of pythons have been captured. Scientists are looking for ways to control this and other invasive organisms.

**A Burmese python fights with an alligator in the Florida Everglades. Pythons now compete with alligators as the top predator in the Everglades.**

## Before You Move On

1. Describe some ways that animals change the environment by making homes.
2. How do nutria harm the environment?
3. **Draw Conclusions** People traveling to different countries are not allowed to take live plants with them. Why?

# People Change the Environment

**Using Natural Resources** Plants and animals change the environment in some major ways. One kind of living thing, however, causes the most change—people.

People change the environment when they use natural resources. These include air, water, land, plants, and animals. People used lots of rocks and other land resources to make the concrete for this dam. How do you think building the dam changed the environment?

This part of the Columbia River became flooded when the Grand Coulee Dam was built.

The Grand Coulee Dam blocked the Columbia River, which backed up to form Lake Roosevelt.

People use water behind the dam to make electricity. Water rushes through tunnels in the dam. The force of the water spins large magnets near wires to make electricity.

The dam has been helpful to people but harmful to other living things. A dam causes the river behind it to back up and flood the land. Plants and animals that lived near the river lost their homes. Dams in the northwestern United States sometimes block salmon from swimming upstream to lay their eggs. So people build fish ladders that go around the dams. The fish ladders help make sure plenty of salmon can live and grow in the wild.

A fish ladder is a chain of pools built like steps so fish can bypass a dam. The fish leap out of one pool and into the next.

**Building Homes and Roads** People often change the environment to meet their needs. Think about building homes and roads. This construction usually harms the environment in some way. People cut down forests to make room for buildings and roads. They use wood to make furniture and paper. Many animals lose their homes when people cut down forests. Trees can grow back, but not as quickly as they are cut down.

**How did building these homes change the environment?**

Some people are trying to cause less harm to the environment by building "green homes." These homes are made with **recycled** materials. To recycle is to use something again. Green homes are often built using recycled tires or wood taken from old buildings. This helps **conserve** natural resources. To conserve is to use something carefully and not waste it. Conserving saves trees and other resources. Conserving makes the environment healthier for all living things.

This green housing development in London, England, uses a lot of recycled materials.

**Farming** People also change the environment when they grow crops and raise animals. For example, some farming methods cause more soil erosion than others. Some methods use more water than others. When a lot of water is taken from rivers and lakes to grow crops, there is less for other plants and animals. Many farmers use chemicals on their crops to kill insects and weeds. These chemicals get into the soil. Rain washes the chemicals, as well as animal wastes, into streams and lakes. The chemicals and wastes can harm plants and animals.

Water that washes off this farm erodes the soil.

Contour farming slows soil erosion on this farm in Missouri.

Some people farm in ways that do less harm to the environment. Some farmers use contour farming to reduce soil erosion. In contour farming, farmers grow crops in rows that run across a hill instead of up and down it. You may have seen signs for organic foods in grocery stores. Organic farmers do not use chemicals to kill pests or grow their crops. Instead, they use natural substances, plants, animals, or other living things to keep their crops healthy.

People at this organic farmers' market are buying fresh tomatoes grown on nearby farms.

**Littering and Pollution** People use many natural resources, but they do not always use them wisely. They often damage the environment by polluting and littering. **Pollution** is any harmful substance that people put into the air, water, and soil. Pollution comes from cars, factories, and many other sources. Litter thrown into rivers, fields, and other places harms plants and animals. It pollutes soil and water. Pollution and litter also can harm people's health.

Oil surrounds this boat in Alaskan waters after an oil spill.

Oil from a spill covers this penguin.

Volunteers clean the oil off this penguin and save its life.

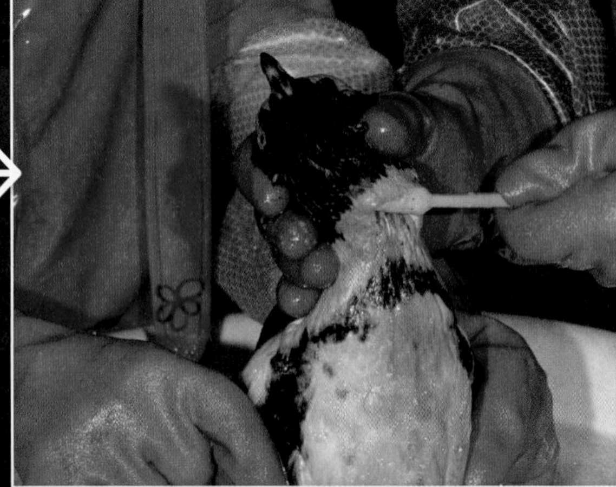

Today, more people understand the ways in which they harm the environment. People are working to stop pollution. Burning coal, oil, and natural gas causes pollution. So scientists are trying to find new ways to run cars and factories. Many cities have set up recycling programs. People get together in groups to pick up litter safely from beaches, rivers, and roads. What can you do to help the environment?

## Science in a Snap! What Can You Recycle?

Make a list of things you use at school or at home that you could recycle. Share your list with your classmates.

Choose one of the items. Make a plan to recycle it.

**How will recycling this item help the environment?**

### Before You Move On

1. What are some ways that people change the environment?
2. How is organic farming different from other kinds of farming?
3. **Apply** Suppose you want to help the environment by changing how you get your food. Describe what you might do.

# USING INVASIVE ORGANISMS TO HELP THE ENVIRONMENT

Invasive plants and animals often damage the environment. Invasive organisms usually spread very quickly, so they are difficult to remove. One way to solve the problem is to put invasive organisms to work! For example, people in Virginia are using invasive plants to feed goats. The goats graze on invasive plants such as Asiatic bittersweet, multiflora rose, and some kinds of honeysuckle.

**Asiatic bittersweet is native to Asia. It is a vine that can pull down trees and crowd out other plants.**

The goats may have to graze different parts of the land a few times before they kill the invasive plants. It might take several years before the plants are gone. In the meantime, the goats have a new source of food. An added bonus: wastes from the goats help improve the soil.

**Goats are known for eating almost anything. Here, they help get rid of invasive plants by grazing on them.**

**Harvesting Invasive Organisms** In the southwestern United States, saltcedar is an invasive plant that crowds out native trees and bushes. Scientists are testing saltcedar to see if it would be useful in building homes. If it is, people could use less native wood to build homes and get rid of an invasive plant all at once!

Saltcedar is native to Europe, Asia, and Africa. People brought it to the United States to add beauty to the land.

Invasive organisms damage underwater environments too. Some invasive fish crowd out native fish by eating them or their food. But what if you could turn invasive fish into food for people?

That is what some people are trying to do with the red lionfish. This fish comes from the western Pacific Ocean. It is an invasive animal in the Florida Keys and the Caribbean Sea. Lionfish damage coral reefs and the organisms that live there. People have started catching red lionfish and serving them at restaurants. It is hoped that as this new source of food becomes more popular, the numbers of lionfish in the Caribbean will start going down.

These poisonous spines protect the red lionfish. Few other fish can eat them.

# Conclusion

Living things interact with their environment in many ways. The different seasons cause changes in plants and animals. Plants change the environment by helping to make soil, slow erosion, clean the air, and cool Earth. Animals change the environment by building homes, helping to make soil, and feeding. Invasive plants and animals damage the environment by using up resources. People change the environment by using natural resources. They harm it by polluting and littering. People also can do things to improve the environment by conserving natural resources, recycling, and stopping pollution.

**Big Idea** Living things interact with their environment by changing the environment and changing in response to it.

## Vocabulary Review

**Match** each of the following terms with the correct definition.

**A. invasive organism**
**B. conserve**
**C. recycle**
**D. pollution**

1. To use resources in a careful way
2. Harmful substances that people put into the air, water, and soil
3. To use something again
4. A plant or animal that does not belong in a place and can harm it

## Big Idea Review

1. **Describe** Describe how seasonal changes in plants are different in different parts of the United States.

2. **Recall** What do some animals do to survive cold weather?

3. **Recall** Explain what building "green homes" has to do with conserving resources.

4. **Compare** How are invasive plants and invasive animals the same?

5. **Predict** Suppose you see a sign that says "New Homes Coming Soon" outside of a forest. What do you think might happen to the forest and the living things in it?

6. **Infer** A river backs up behind a dam and forms a lake. What do you think happens to the river on the other side of the dam? How does this affect the plants and animals downstream from the dam?

## Write About Helping the Environment

**Evaluate** Look at the picture. How does choosing organic foods help the environment? What can you do to help the environment?

# Want to help care for our planet?

## Conservationist Severn Cullis-Suzuki leads the way.

**What do you do as a conservationist?**

I try to tell people that we need to take care of the Earth. I am trying to help people see that even though we depend on the natural world, we are hurting it with the way we live. I think we need to change the way we live.

**What did you study in school?**

I earned a degree in biology in college. Then I earned a Master of Science in Ethnoecology, which is learning about indigenous people's science and their views on taking care of the Earth.

**What is a work day like for you?**

I do all kinds of different things on my computer at home. I write articles and speeches and do research. I travel, do public speaking, and work with the media. Just recently I filmed a TV series about water.

TECHTREK
myNGconnect.com

Digital Library

Conservationist Severn Cullis-Suzuki grew up in British Columbia, Canada, where she learned to love the natural world. Today, she works to protect it.

Cullis-Suzuki speaks to groups of people to spread the word about caring for our planet.

**What's been your greatest accomplishment so far?**

When I was a teenager, I was invited by the United Nations to work with people from all over the world to put together the Earth Charter. The charter is a code of conduct for the way people should act toward the Earth. I am very proud of the Earth Charter.

**What advice would you give young people who are interested in doing what you do?**

I think a love of the Earth is the most important thing you need if you intend to fight for it. I always give the same advice to any young person who is interested in working for change in the world: Follow your heart. You will be most successful if you are doing what you are excited about and are good at.

**Where do you hope your job has made the biggest impact?**

I like to think that what I have done has helped youth voices be heard and that they have helped people take action in their communities. Now, more than ever, we need the voices of young people to remind the world of what is truly important.

NATIONAL GEOGRAPHIC

# BECOME AN EXPERT

## Green Movements Around the World

Have you ever been to an Earth Day party? At one of these parties, you might have planted a tree or helped clean up your neighborhood. Earth Day is April 22. It is a day to appreciate the environment and to understand how important it is to protect it. Over the last 50 years, many people have started to realize that we have been accidentally harming the environment. They want to stop the damage and start fixing the damage that has been done. This desire to take action to help the environment is called a green movement.

**Students often gather in groups to clean up their communities—on Earth Day or any other day.**

TECHTREK
myNGconnect.com

Student eEdition

Digital Library

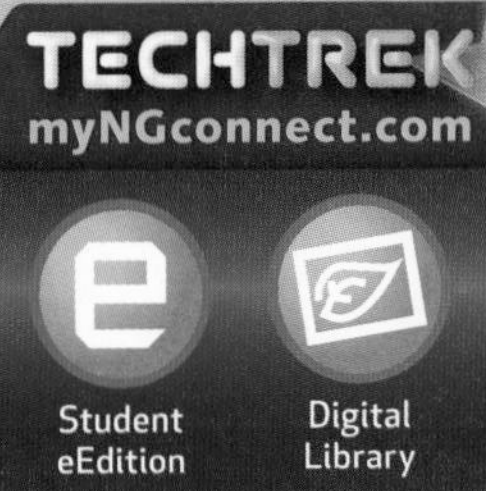

All over the world, people have started groups that work to protect the environment. Some groups focus on protecting forests. Others try to protect animals and the places where they live. Some work to pass laws to keep the air and water clean and to fight global warming. Still others try to keep **invasive organisms** from crowding out native plants and animals. On the next pages, you will learn more about some of these environmental groups.

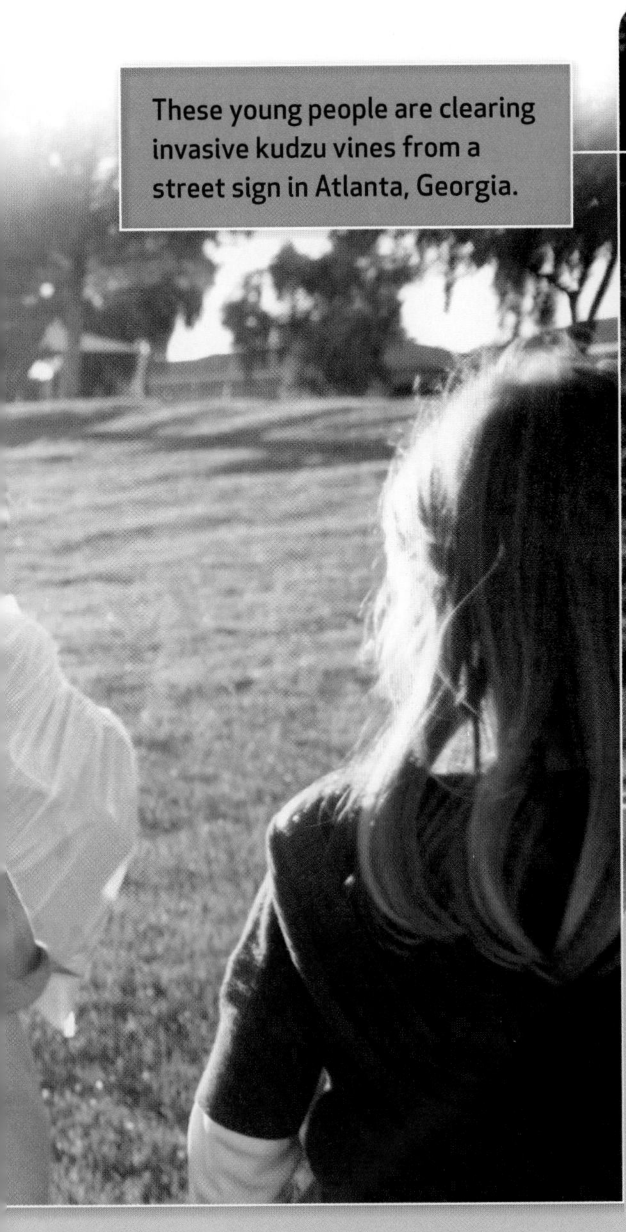

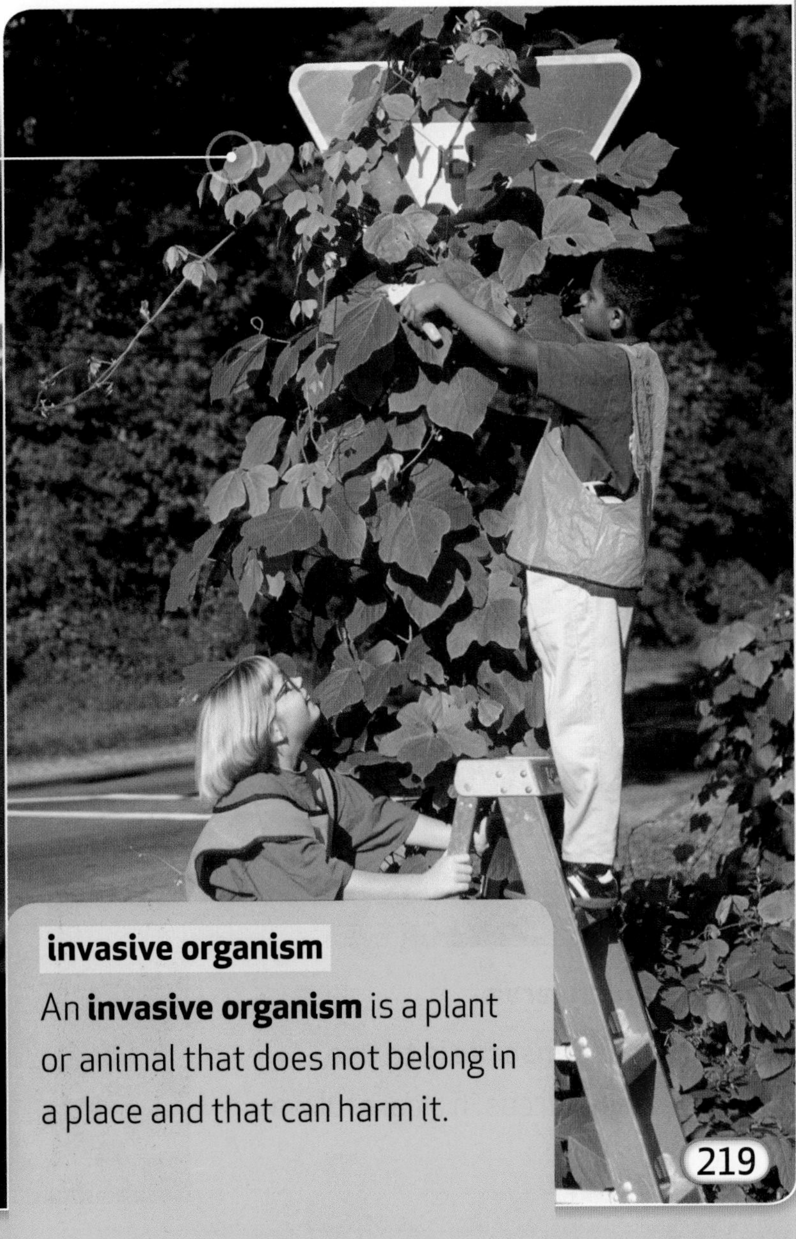

These young people are clearing invasive kudzu vines from a street sign in Atlanta, Georgia.

**invasive organism**

An **invasive organism** is a plant or animal that does not belong in a place and that can harm it.

**Green Belt Movement** "She thinks globally and acts locally," said the chair of the committee who awarded Dr. Wangari Maathai the 2004 Nobel Peace Prize. Maathai won the prize for her work with the Green Belt Movement (GBM). She started the GBM in 1977. The GBM works to **conserve** Africa's trees. Trees slow soil erosion. They also provide places for animals and plants to live. Because of the work of the GBM, more than 40 million trees have been planted in Africa!

Wangari Maathai of Kenya, Africa, is known all over the world for getting people together for good causes, such as planting trees.

**conserve**

To **conserve** is to use resources in a careful way.

The GBM teaches local women how to plant tree seeds and care for young trees. When the trees are big enough, the women plant them on people's farms and near their homes. Then their neighbors care for the trees. The trees give the people fruit to eat, food for their farm animals, and wood for fires and for building houses. The trees also help make new, healthier soil, which helps farmers. The GBM has shown the people of Africa that trees are good for the environment and for their own lives.

Women gather for a meeting of the Green Belt Movement.

Each of these containers holds a young tree that members of the GBM will plant.

**National Audubon Society** The National Audubon Society (NAS) is one of the oldest environmental groups in the world. One of the goals of the NAS is to protect birds and the places where they live. The NAS owns many wildlife sanctuaries, where people are not allowed to build anything. For example, the Florida Coastal Islands Sanctuary protects birds, such as herons, that build their nests on islands in Tampa Bay. People cannot build on the islands. Birds are safe there.

This egret finds food and shelter among these red mangrove trees.

TECHTREK
myNGconnect.com

Digital Library

These white pelicans and the smaller egret have nothing to fear from people, thanks to those who started the Florida Coastal Islands Sanctuary.

The NAS also helps protect other animals and the places where they live. It works to clean up ecosystems harmed by **pollution**. For example, people used to use a dangerous chemical called DDT to kill insects. But DDT also got into the water and plants and killed birds and other animals. The NAS worked to help ban DDT from use in the United States. That has made the environment safer for animals as well as people.

This healthy eagle chick rests after hatching. About 30 years ago, however, it might not have made it. DDT in birds' bodies made their eggshells so thin that the eggs broke before the baby birds were ready to hatch.

**pollution**

**Pollution** is a harmful substance that people put into the air, the water, and the soil.

**Wildlife Protection Society of India** The Wildlife Protection Society of India (WPSI) helps Indian state governments train forest and police officers to catch people who hurt wild animals. Some people catch and sell wild animals—including monkeys, birds, and reptiles—for pets or to circuses. This is against the law. Ending poaching is another goal of WPSI. Poaching is killing animals in order to sell their hides, fur, or other body parts. Tigers are often poached in India. Poaching is against the law.

The Wildlife Protection Society of India helps protect animals, such as Bengal tigers, from poachers.

The WPSI also works to help other animals in India, such as otters, sea turtles, and elephants. Pollution and construction have damaged or destroyed many of the places these animals used to live. For example, people have built buildings and iron mines on many elephant paths. Elephants used these paths to move around the country. The WPSI has studied ways to make new paths or bring back the old ones. Businesses and state governments are using their ideas to help elephants.

**It's not a parade. It's a herd of elephants heading along a wide path to a river for a drink. The WPSI helps to keep these paths open, or make new ones.**

**Movement of Small Farmers** In Brazil, the Movement of Small Farmers (Movimiento de Pequeños Agricultores), or the MPA, is trying to stop the spread of very large farms owned by companies. Large areas of rainforest are often cleared to make room for large company-owned farms. When rainforests are cut down, soil erodes. Plants and animals lose their homes. The MPA is working for a future in which Brazilian land is farmed by local farmers, not by large companies.

Huge areas of rainforest in Brazil have been cleared to make room for large farms and cattle grazing.

Like the MPA, people in all countries can do things to help the environment. They can do something small, such as convince their family to **recycle**. They can do something large, such as start a group like the MPA. Either way, they are making the world a better place.

School and community gardens have had a positive impact on the environment. Here, a teacher and her students pull weeds from a school garden.

**recycle**

To **recycle** is to use something again.

NATIONAL GEOGRAPHIC

# CHAPTER 5 SHARE AND COMPARE

**Turn and Talk** How do different groups around the world help protect the environment? Form a complete answer to this question together with a partner.

**Read** Select two pages in this section that are the most interesting to you. Practice reading the pages so that you can read them smoothly. Then read them aloud to a partner or small group. Talk about why the pages are interesting.

my SCIENCE notebook

**Write** Write a conclusion that summarizes what you have learned about groups that help the environment. State what you think is the Big Idea of this section. Share what you wrote with a classmate. Compare your conclusions. Did you recall some of the specific things that different groups in the green movement do?

my SCIENCE notebook

**Draw** Draw a picture of what you think a National Audubon Society sanctuary for birds might look like. Combine ideas from your drawing with those of your classmates to make a new drawing that includes the best ideas.

## After reading Chapter 6, you will be able to:

- Explain that humans and other living things have different body parts that help them get what they need to survive. **CIRCULATORY AND RESPIRATORY SYSTEMS, DIGESTIVE SYSTEM, SKELETAL, MUSCULAR, AND NERVOUS SYSTEMS**
- Recognize that animals are composed of different parts performing different functions but working together for the well-being of the organism. **CIRCULATORY AND RESPIRATORY SYSTEMS, DIGESTIVE SYSTEM, SKELETAL, MUSCULAR, AND NERVOUS SYSTEMS**
- Describe the major organs and function of different body systems, including the circulatory, respiratory, digestive, skeletal, muscular, and nervous systems. **CIRCULATORY AND RESPIRATORY SYSTEMS, DIGESTIVE SYSTEM, SKELETAL, MUSCULAR, AND NERVOUS SYSTEMS**
- Recognize that healthful habits include eating a balanced diet, getting regular exercise, using proper hygiene, and avoiding drugs, alcohol, and tobacco. **KEEPING BODY SYSTEMS HEALTHY**
- Science in a Snap! Describe the major organs and function of different body systems, including the circulatory, respiratory, digestive, skeletal, muscular, and nervous systems. **CIRCULATORY AND RESPIRATORY SYSTEMS**

CHAPTER

# 6

# HOW DO THE PARTS OF AN ORGANI

**You can whirl** across a stage, climb a mountain, or walk a dog. Have you ever wondered how your body works? Like the bodies of all living things, your body needs certain things not only to survive but also to thrive. Some parts of your body help you get oxygen, food, and water. Other parts help you respond to changes.

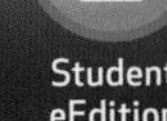

Student eEdition | Vocabulary Games | Digital Library | Enrichment Activities

# SM WORK TOGETHER?

All the parts of this young woman's body are working hard to help her perform at her peak.

# SCIENCE VOCABULARY

## **organ** (OR-gun)

An **organ** is a body part that does a specific job in an organism. (p. 234)

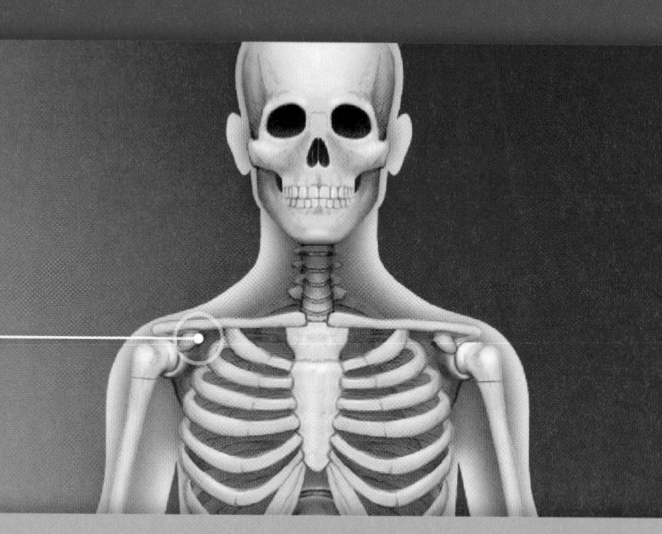

Each bone in a skeleton is an organ.

## **system** (SIS-tum)

A **system** is a group of organs that works together in an organism to help it live and grow. (p. 235)

The children use many different body systems to ride their bikes.

## **circulatory system** (SIR-kyū-luh-tor-ē SIS-tum)

The **circulatory system** is a group of organs that carries blood throughout the body. (p. 236)

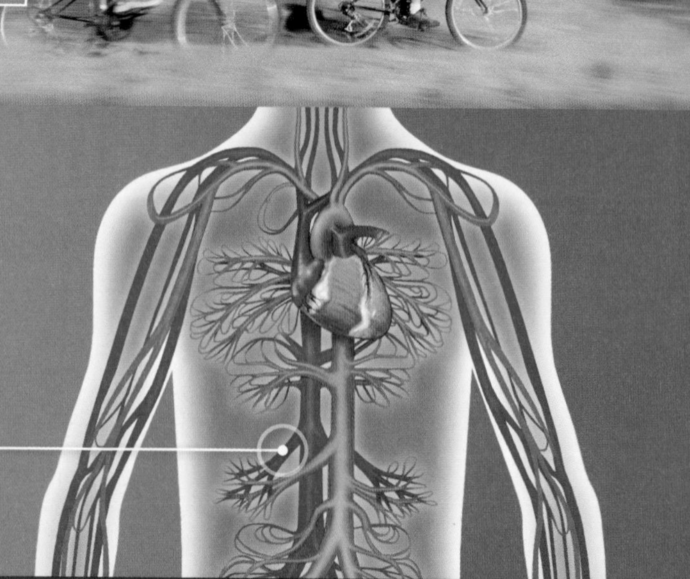

The blood vessels in the circulatory system reach every part of the body.

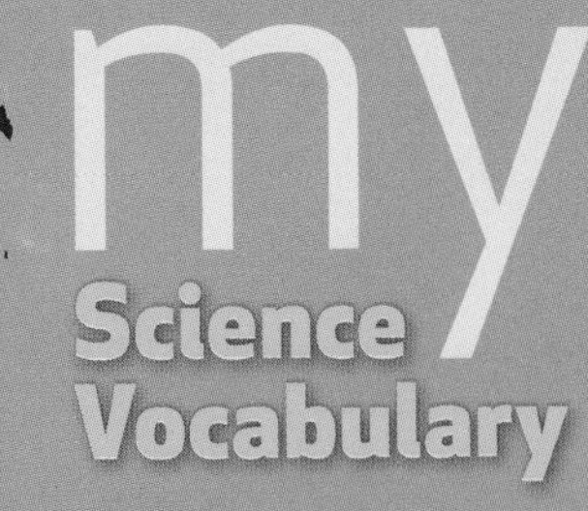


**circulatory system**
(SIR-kyū-luh-tor-ē SIS-tum)

**digestive system**
(dī-JES-tiv SIS-tum)

**nervous system**
(NER-vus SIS-tum)

**organ**
(OR-gun)

**respiratory system**
(RES-pur-uh-tor-ē SIS-tum)

**system**
(SIS-tum)

## respiratory system
(RES-pur-uh-tor-ē SIS-tum)

The **respiratory system** is a group of organs that brings oxygen into the body and gets rid of carbon dioxide. (p. 239)

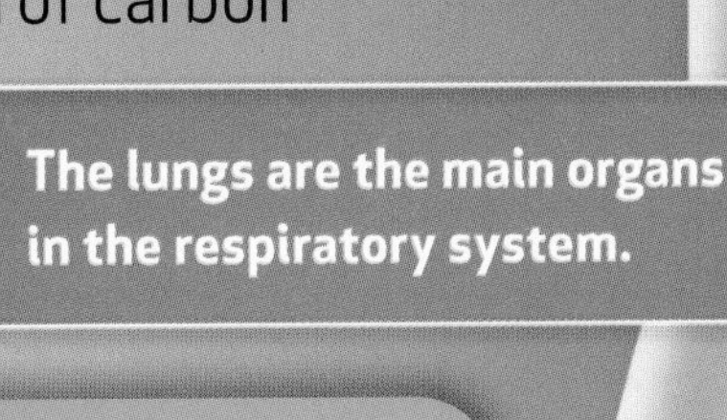
The lungs are the main organs in the respiratory system.

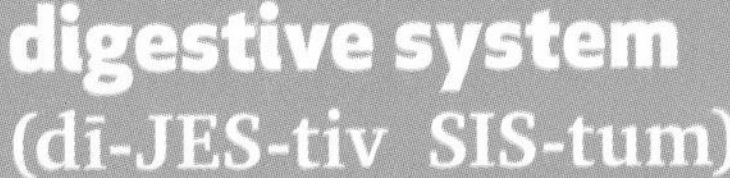

## digestive system
(dī-JES-tiv SIS-tum)

The **digestive system** is a group of organs that breaks down food into energy and materials that the body can use. (p. 240)

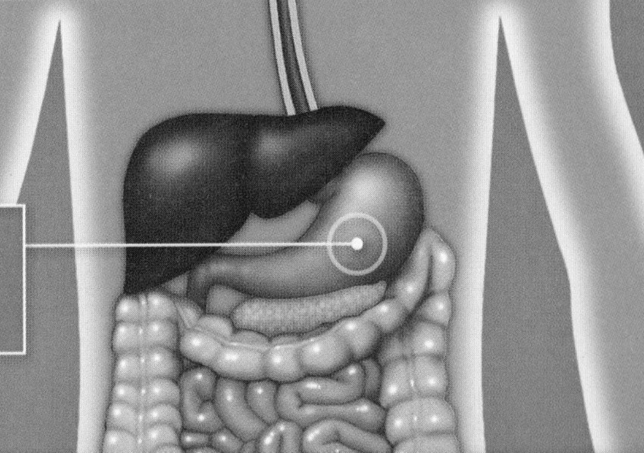
The stomach is part of the digestive system.

## nervous system
(NER-vus SIS-tum)

The **nervous system** works as a communication network to help the body respond to its environment. (p. 246)

Information moves quickly through the nervous system from nerves to the brain.

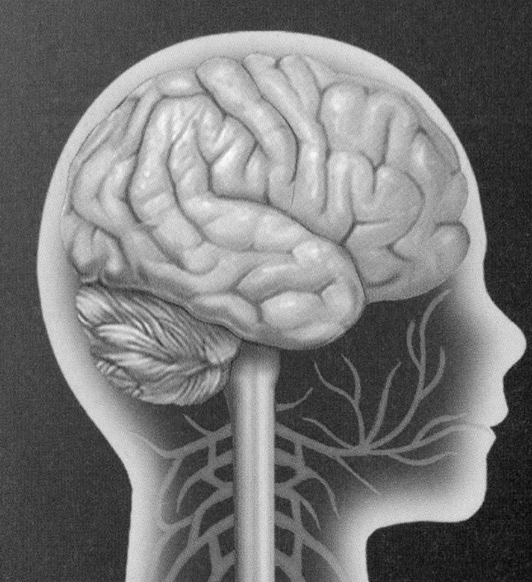

# Circulatory and Respiratory Systems

When you ride a bicycle, you turn the pedals. The pedals turn a gear, which moves a chain. As the chain turns, it moves another gear at the back wheel. Each part has a special job to do. When all of the parts work together, the bike moves forward.

You have body parts that work together too. These body parts are called **organs**. An organ is a structure that does a specific job in an organism. The human body has many organs.

Bicycles have parts that work together just like your body does.

A system is a group of organs that works together in an organism to help it live and grow. You have many systems in your body. For example, one system helps you get oxygen from the air. Another helps the oxygen get to where it is needed in your body. Some body systems help people stretch and bend. Other body systems help people get energy from the food they eat. All these body systems must work together to keep people healthy and help them do the things they want to do.

The pedals, chain, and gears are like organs. Each has a specific job to do. Together, they make up a system that moves the bicycle.

**Circulatory System** Have you ever felt your heart pounding while running a race? Your heart is the main organ in your **circulatory system.** This system carries blood throughout your body. Each time your heart beats, it pumps blood. The blood travels around the body through blood vessels.

Each part of your body needs oxygen to do its job. Your body takes the oxygen it needs from the blood. Blood, rich with oxygen, circles through the body in an endless loop.

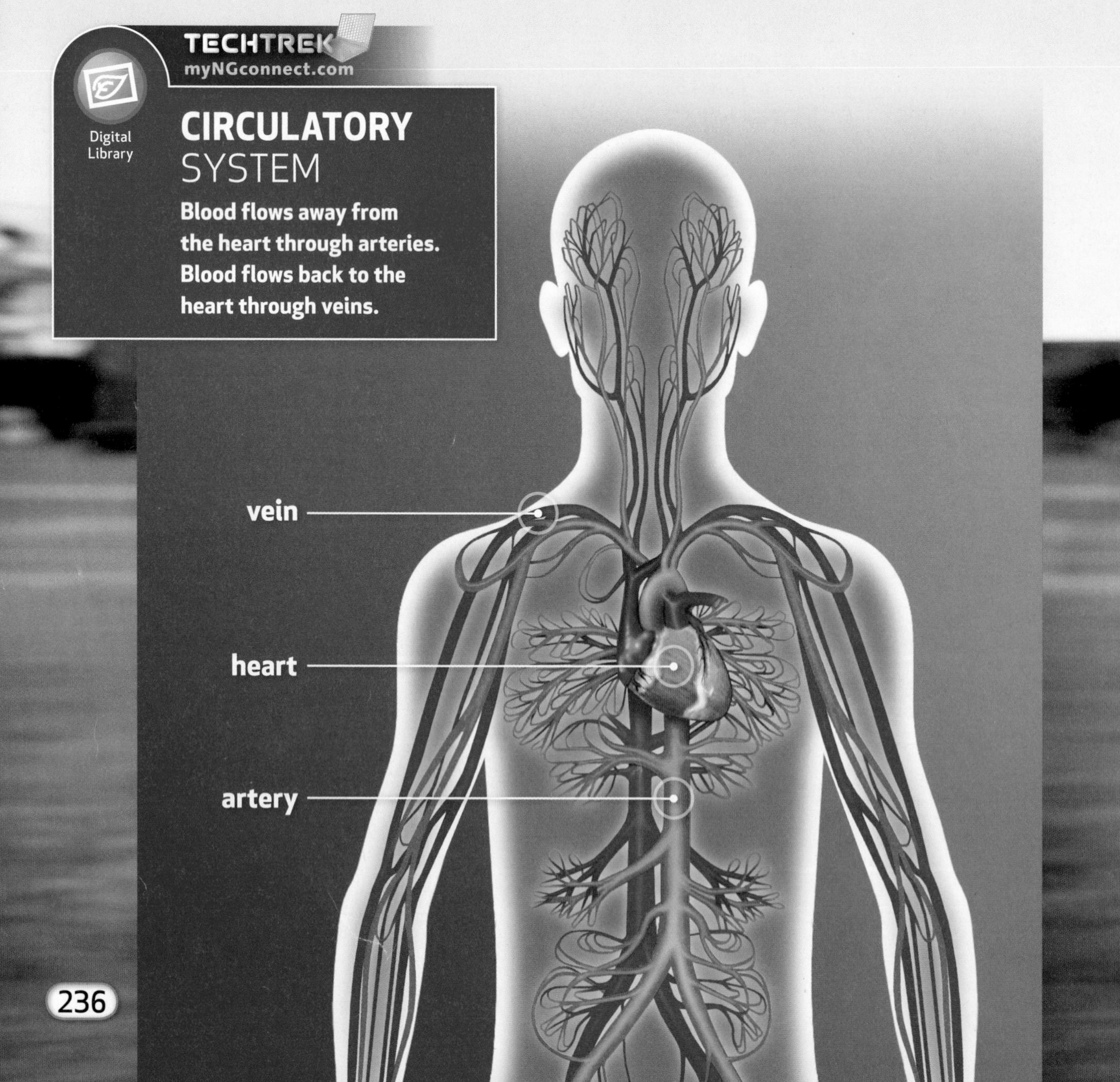

Blood vessels that carry blood to the heart are called veins. Blood vessels that carry blood away from the heart are called arteries. The arteries pulse, or move, in time with the beating heart. One place to feel this pulse is in your wrist. If you can feel your pulse, you can tell how fast your heart is beating.

## Science in a Snap! Taking a Pulse

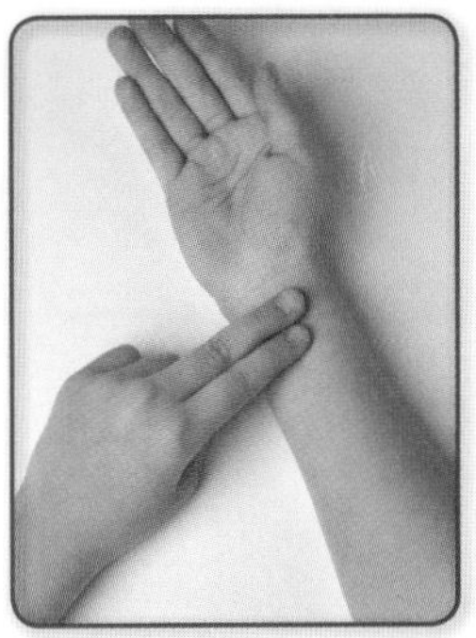

Find your pulse by placing two fingers of one hand on the inner wrist of the other hand.

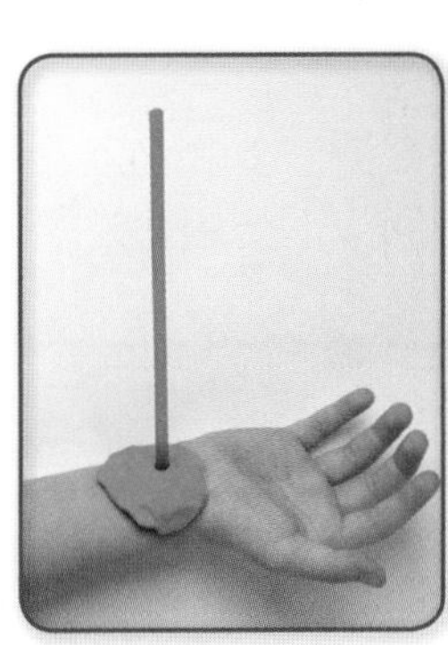

Place a piece of clay over the spot where you felt the pulse. Carefully put a straw into the clay. Count how many times the straw moves in one minute.

**How can you make the straw move more quickly?**

**The harder the body works, the faster the heart pumps. This helps get more oxygen to all the parts of the body.**

**Respiratory System** Humans need oxygen to live. Your body uses oxygen to get energy from the food you eat. Oxygen helps your body do the work it needs to do.

These soccer players' respiratory systems help them get the oxygen they need.

The respiratory system brings oxygen into the body and moves carbon dioxide out. When you breathe in, your lungs take oxygen from the air. The heart pumps blood to the lungs, and the blood takes in some of that oxygen. This blood flows through the circulatory system, carrying oxygen to all the parts of your body. The blood absorbs the carbon dioxide, which the body does not use, and carries it back to the lungs. When you breathe out, the carbon dioxide leaves your body.

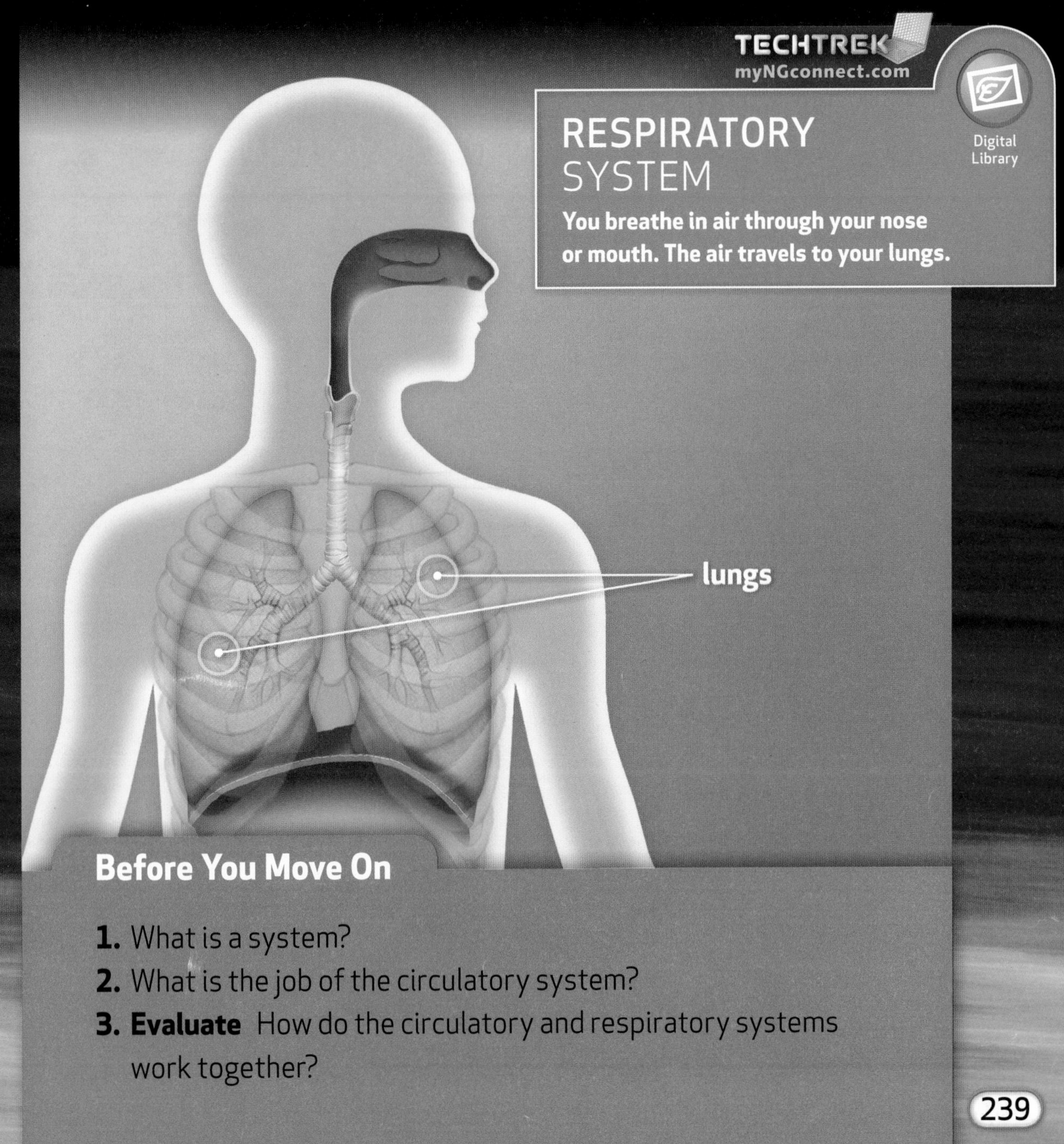


## RESPIRATORY SYSTEM

**You breathe in air through your nose or mouth. The air travels to your lungs.**

## Before You Move On

1. What is a system?
2. What is the job of the circulatory system?
3. **Evaluate** How do the circulatory and respiratory systems work together?

# Digestive System

What kinds of foods do you like to eat? Whether you are eating a salad or a pizza, your body is able to change that food into energy. The **digestive system** breaks down food into energy and materials that the body can use.

The girl in the picture is eating. First, she chews the food, breaking it up into smaller pieces. Her saliva breaks down the pieces even further. When she swallows, the pieces go down a tube called the esophagus, and then into the stomach.

The stomach churns the pieces, breaking them down even more. From the stomach, the digested food passes into the intestines. Now the intestines can absorb materials from the food that the whole body will use.

Eating is just the first step in getting energy from food.

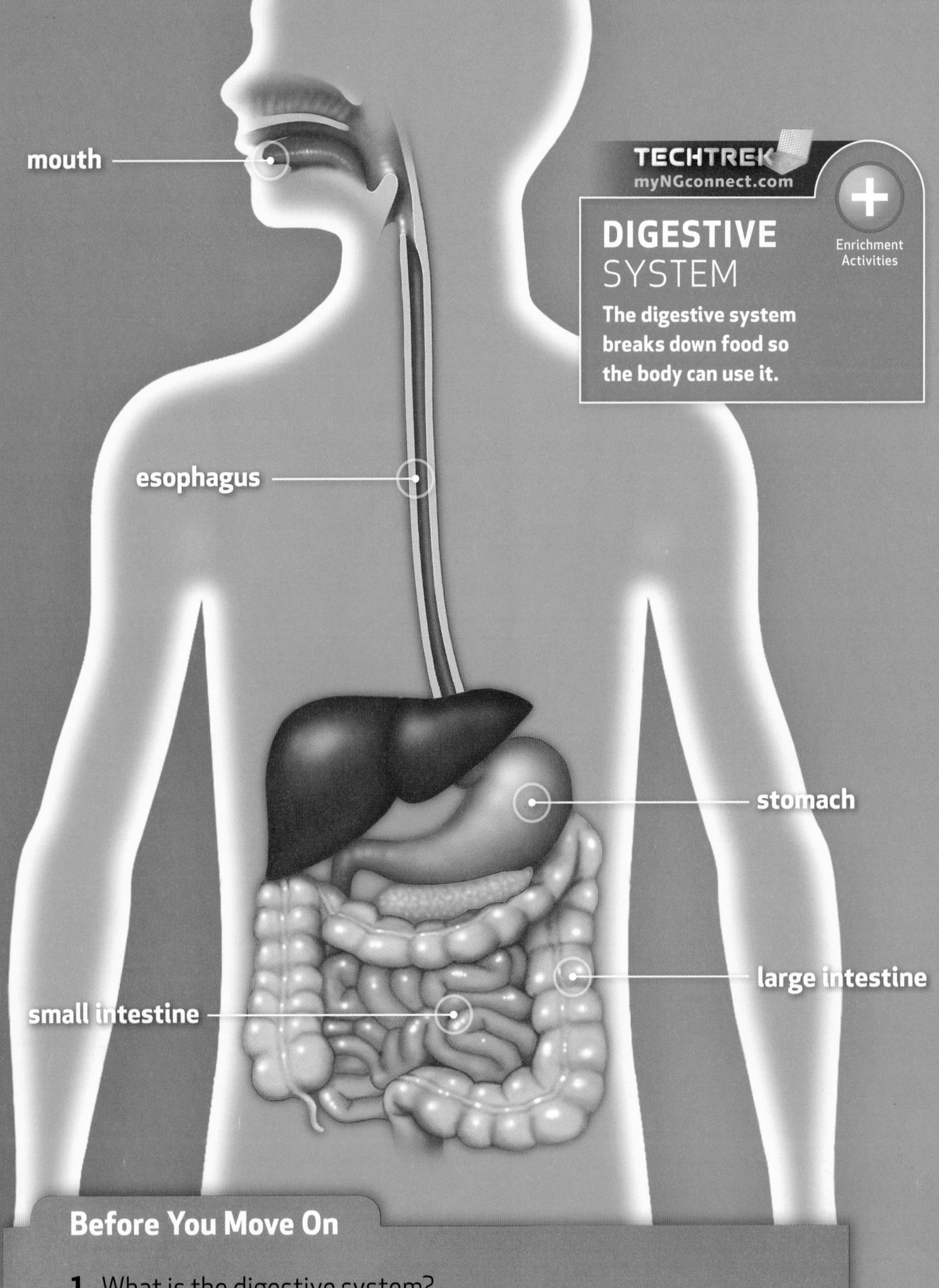


## Before You Move On

1. What is the digestive system?
2. What is the stomach's role in the digestive system?
3. **Draw Conclusions** How does the digestive system work together with other body systems to keep your body healthy?

# Skeletal, Muscular, and Nervous Systems

**Skeletal System** Hundreds of bones make up your skeletal system. The skeletal system supports and gives shape to the body. In many places in your body, two bones come together and form a joint. Your knees, shoulders, and other joints allow you to bend and move in different ways.

Some bones protect soft organs. For example, the ribs form a protective cage around the organs in the chest.

Another part of the skeletal system, the backbone, is made up of many small bones. The backbone supports your body and allows you to bend and twist in many ways.

These children use their skeletal system to bend and move.

## SKELETAL SYSTEM

**Your skeletal system supports and protects your body.**

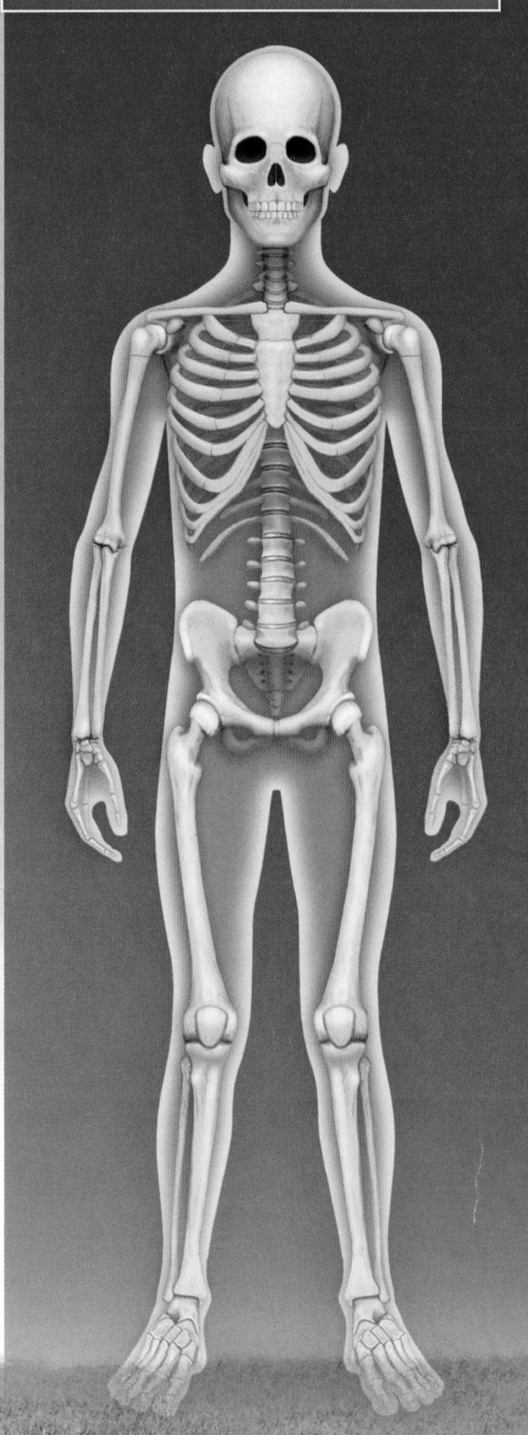

**Muscular System** Your muscular system is made up of more than 600 muscles. You use those muscles when you make large movements, such as pushing open a door, and when you make small movements, such as snapping your fingers. Your muscles move your body parts. Look at the diagram below. Some of your muscles are attached to bones. When the muscle contracts, it gets shorter and thicker and pulls on the bone. The bone moves. You can control the muscles that are connected to bones.

## ARM **MUSCLES**

**The muscles in your arms help you extend and contract your arms.**

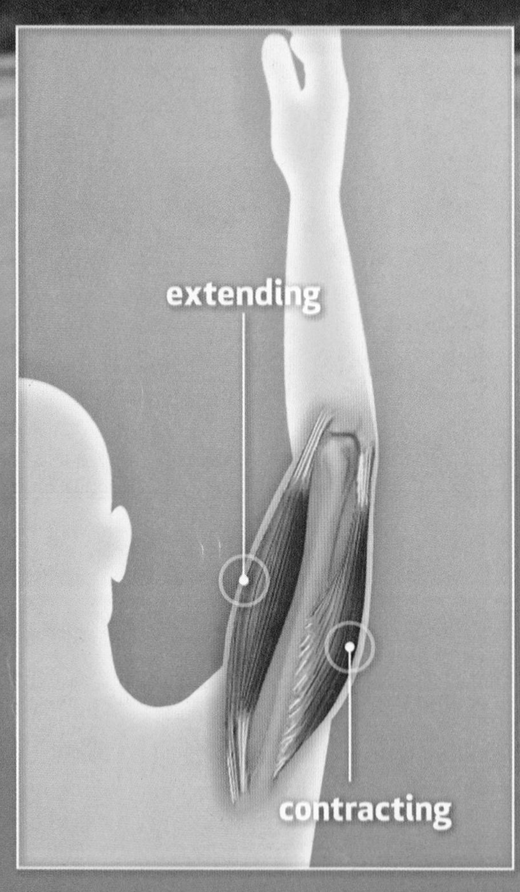


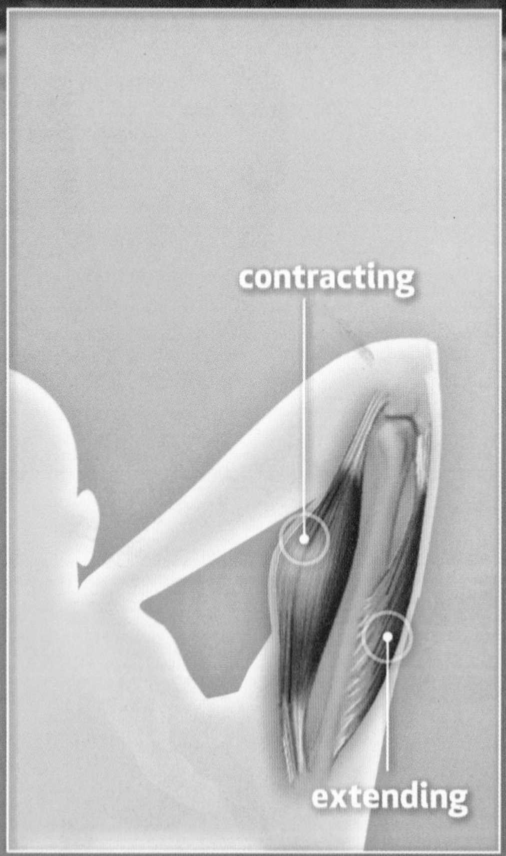

There are other muscles in the body that you do not control. For example, heart muscles keep pumping blood throughout your body. You don't have to think about moving these muscles. They move automatically.

Muscles let you make many facial expressions.

This boy's muscles help him move the paddle through the water.

**Nervous System** When you touch something really hot, you pull your hand away quickly. How does your body know when something is too hot to touch? The **nervous system** is the body's communication system. Nerves gather information about the environment and send signals through the spinal cord to the brain. When you touch something really hot, nerves send messages to the brain telling about the object. Your brain then sends a signal to your hand telling it to pull away from the hot object.

TECHTREK
myNGconnect.com
Digital Library

**The brain is the most important organ in the nervous system.**

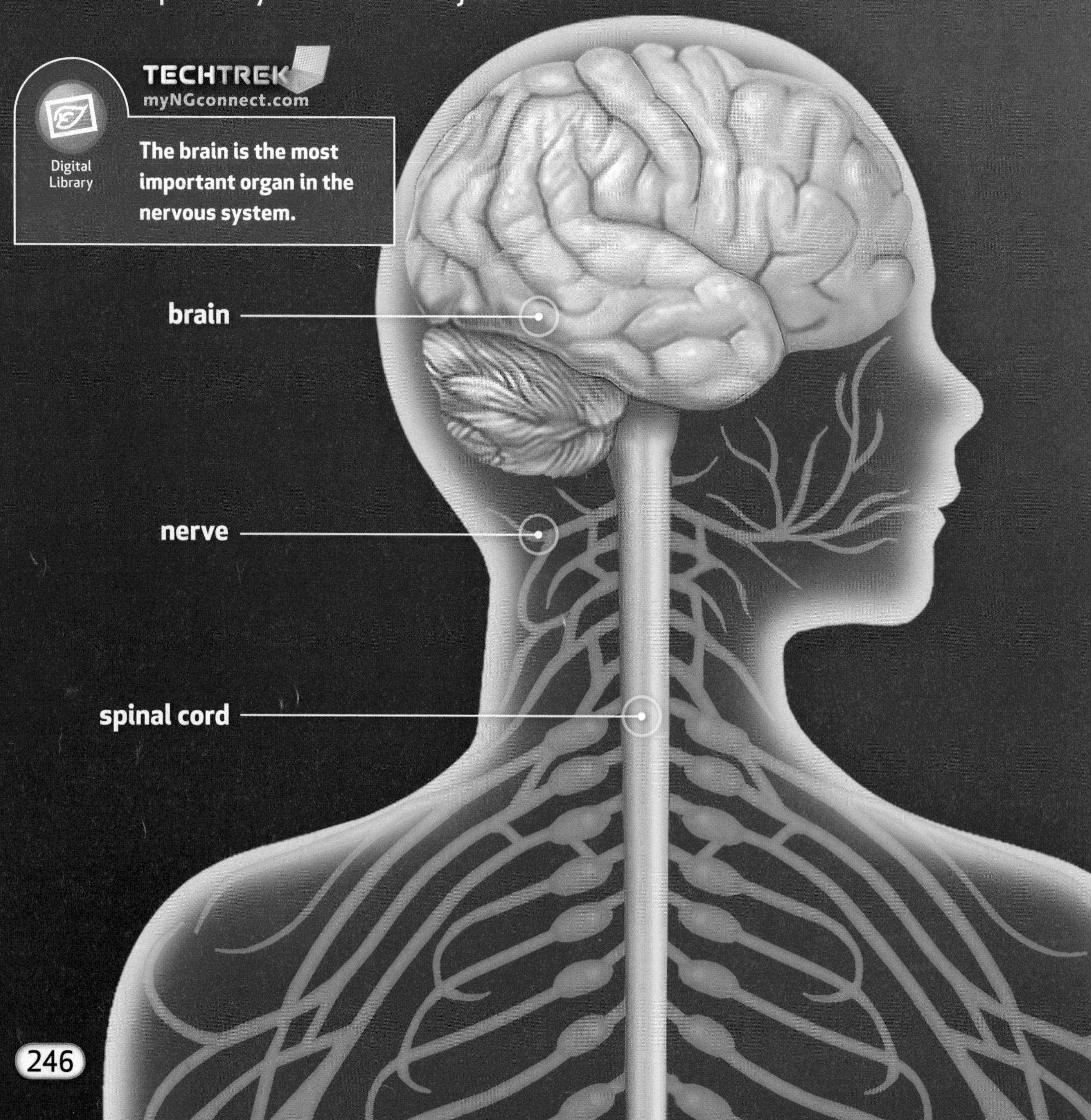

The brain is like a control center. It gets messages about all your senses—sight, hearing, touch, taste, and smell. The brain uses the messages to tell your body how to respond to changes in your environment. For example, when you get hot, your brain tells your body to perspire. The perspiration cools your body.

**The snowball feels cold. A network of nerves spread throughout the body provides detailed information about the snowball to the brain.**

## Before You Move On

1. What is the skeletal system?
2. What is the brain's role in the nervous system?
3. **Infer** Suppose you go to a hot beach. How might your nervous system help your body respond?

# Keeping Body Systems Healthy

**Diet, Exercise, and Rest** What can you do to keep your body systems working the way they should? One of the best things to do is to eat many different kinds of food. Eating a balanced diet can help you get all the nutrients your body needs. Nutrients are materials that help organisms live and grow. Eating healthful foods helps your digestive system work well. It also helps keep your circulatory system free of unhealthful materials that can clog blood vessels.

**Most experts say young people should get at least an hour of exercise a day. You might walk, ride a bike, or play basketball.**

What are your favorite activities? Getting regular exercise is a health habit that helps all your body systems stay strong. It keeps your lungs healthy. That helps your respiratory system. It keeps your heart healthy. That helps your heart pump blood through your body. Regular exercise can make your bones and muscles stronger too.

In addition to exercise, your body needs rest to stay healthy and grow. Most ten-year-olds need about ten hours of sleep a night. Getting enough rest helps all your body systems work well.

A balanced meal contains many nutrients that people need to stay healthy.

**Healthful Habits** Look at your hands. Do they look clean? You might not think so if you could see them under a microscope!

When people touch something, such as a book or a doorknob, they often leave some germs behind. When you touch that book or doorknob, you pick up the germs the other people left on it. Germs can cause you to become ill. To keep from getting sick, wash your hands before you eat and after using the restroom. Washing your hands often is the best way to protect yourself from harmful germs.

**Your hands can be loaded with germs and bacteria. The bacteria shown below are magnified many times.**

**Washing your hands can keep you from getting sick and from passing your germs to other people.**

Staying away from things that can harm your body also keeps your body healthy. These things include drugs, alcohol, and tobacco. They could cause diseases in many of your body systems, and may temporarily alter your thinking and behavior.

**Healthful habits keep these kids happy and growing strong.**

## Before You Move On

1. Why is eating a balanced diet important?
2. How can washing your hands help protect you from getting a cold?
3. **Draw Conclusions** A doctor advises a patient who does not exercise to find a physical activity she likes to do. Why might the doctor have asked her to choose her own activity?

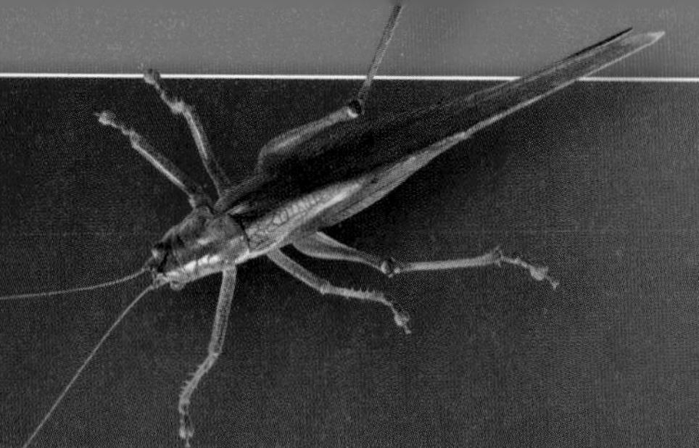

# THE BODY SYSTEMS OF ANIMALS

Animals have body systems that do many of the same jobs as human systems. Like humans, the grasshopper has a circulatory system. The grasshopper's blood carries nutrients throughout the grasshopper's body. The blood also carries away waste. In this way, the grasshopper's circulatory system is similar to a human's circulatory system.

Grasshopper blood does not flow through a network of blood vessels. The heart pumps the blood through a tube to open spaces in the body.

A horse's heart and lungs look like those same organs in humans. They work the same way too. The horse's respiratory system gets the oxygen, and the circulatory system carries it throughout the body. These systems work together in the horse's body just like they do in your body.

Galloping horses need energy and oxygen just like human runners do.

Many animals have a skeleton to support their body, protect the soft parts, and allow the body to move. Some animals have a backbone. Their skeleton is inside their body. Horses, cats, and dogs have a backbone. Humans do too. Some animals, such as lobsters, have a skeleton on the outside of the body. A lobster's tough shell is different from your skeletal system, but it does the same job. It gives the lobster its structure and protects it.

**Unlike human skeletons, a lobster's skeleton does not grow as the lobster grows. Instead, it splits open and the lobster forms a new, bigger skeleton around itself.**

Nervous systems help animals survive in their environment. Lobsters use their eyes, antennae, and hair-like parts on their shells to gather information underwater. This information is sent along the nerve cord to the brain.

In animals with backbones, information passes from nerves to the spinal cord to the brain. In this way, birds, dogs, and humans are alike. They all see through their eyes, hear through their ears, and smell through their nose.

Hawks and other birds of prey use their sharp eyesight to gather information. They can detect small animals moving along the ground, even when the birds are soaring high above.

People have many body systems that do certain jobs. The respiratory system takes in oxygen. Then the circulatory system moves the oxygen-rich blood throughout the body. The digestive system breaks down food. The skeletal system gives the body structure and protects it, while the muscular system helps the body to move. The nervous system helps the body respond to its environment. Healthful habits help people keep their body systems healthy.

**Big Idea** The parts of an organism work together in systems. These systems help the body get oxygen and nutrients it needs. Systems also help organisms respond to their environment.

## Vocabulary Review

**Match** each of the following terms with the correct definition.

**A. system**
**B. nervous system**
**C. organ**
**D. digestive system**
**E. respiratory system**
**F. circulatory system**

1. A group of organs that brings oxygen into the body and gets rid of carbon dioxide
2. A group of organs that carries blood throughout the body
3. A group of organs that works as a communication network to help the body respond to its environment
4. A group of organs that works together in an organism to help it live and grow
5. A group of organs that breaks down food into energy and materials that the body can use
6. A body part that does a specific job in an organism

## Big Idea Review

1. **Identify** What is the main job of the muscular system?
2. **Name** Which two body systems help your body parts get the oxygen they need?
3. **Explain** Why is it a healthful habit to wash your hands often?
4. **Relate** How do the skeletal and muscular systems work together?
5. **Evaluate** Suppose your brain could not understand nerve signals for pain. Why would that be dangerous?
6. **Apply** Zoos often remind visitors not to feed the animals any popcorn, peanuts, or other food. Why might zookeepers give this warning?

## Write About Human Body Systems

**Summarize** How are each person's body systems working together to allow for this kind of movement?

## Dr. Aaron Snead, Pharmacologist

**What is a pharmacologist's job?**

A pharmacologist's job is to use known medicines and chemicals to change how the body parts of different organisms work. The goal is to gain a better understanding of how living things function.

**What do you remember liking about science when you were in elementary school?**

The thing I liked most about science when I was young was the idea that there was more than one right way to be successful. Since the goal of science is to learn new things, I liked being challenged to come up with new ways to solve old problems.

Dr. Aaron Snead works in his laboratory to study how the body parts of organisms function.

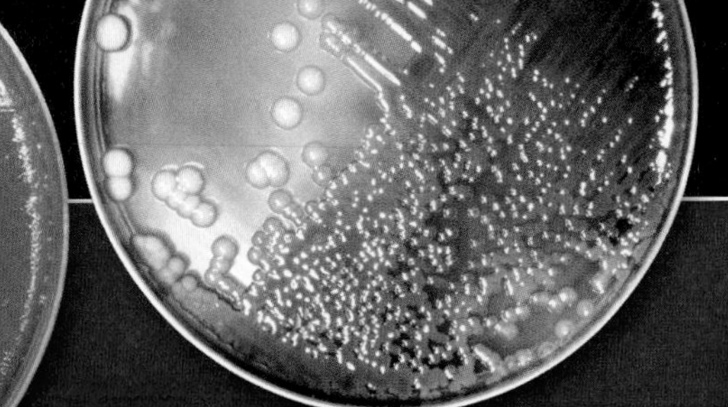

Student eEdition

Digital Library

To be successful in science you need to have curiosity and willingness to follow clues to the end. I think of science as kind of like solving a big jigsaw puzzle. Everyone wants to know what the puzzle will look like, but the skills that make someone patient and observant enough to solve the puzzle are the same skills that make someone a good scientist.

**When you were younger, did you ever see yourself doing what you do now?**

When I was young I loved animals. My family had many different kinds of pets, so I thought that I wanted to become a veterinarian or a doctor. As I got older and learned more about other possible careers in medicine, I realized that I was more interested in understanding disease and studying ways to find cures.

**Where do you hope your job will make the biggest impact?**

I would love to play a key role in the development of a new medicine. I would like to know that my work with those new medicines helps people live healthier, happier lives.

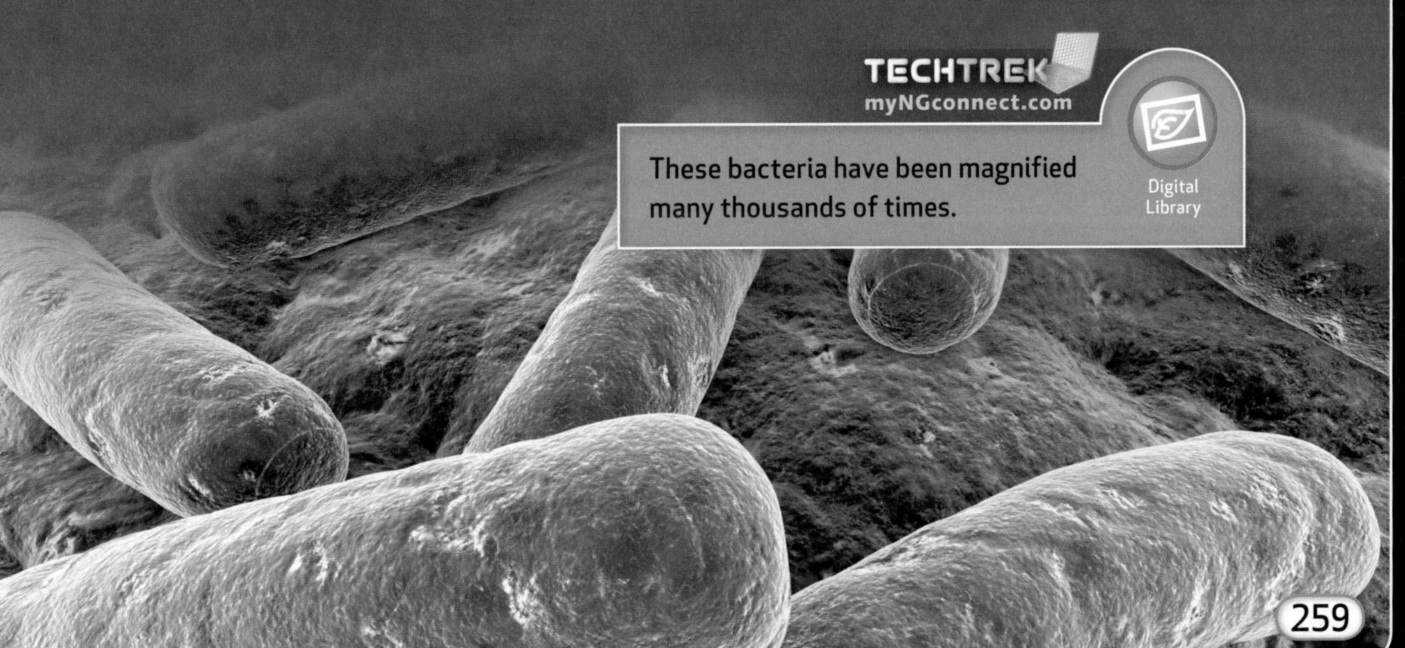


These bacteria have been magnified many thousands of times.

NATIONAL GEOGRAPHIC

# BECOME AN EXPERT

## Technology and Body Systems

When all its **organs** and **systems** are working well, the human body can do some pretty amazing things. But what happens when a part is not working well? That's when scientists and engineers can do some pretty amazing things. Today, doctors use new technologies and materials to replace body parts, help systems work, or find out what is wrong with a patient.

Prosthetic legs may include working knee joints.

**organ**

An **organ** is a body part that does a specific job in an organism.

**system**

A **system** is a group of organs that works together in an organism to help it live and grow.

## A Leg Up

Many years ago, when people lost a leg, they may have used a prosthetic, or artificial, leg made from wood or steel. These legs allowed people to walk, but they were heavy and awkward. Today, many prosthetic legs are made from high-tech materials. Some, made with silicone, look very lifelike. Others, such as the sprinting leg shown here, are made from graphite. This strong, light material supports the body almost as well as bone does. The curved shape also gives the leg some spring, as a muscle does.

This runner's prosthetic leg helps him run.

Prosthetic legs allow world-class athletes to climb mountains and compete in races.

## Nerve Signals

If you want to move your arm, your **nervous system** gives a command. The arm moves. A "bionic" arm, such as the one shown here, works with a person's nervous system to do the same thing. When a person loses an arm, working nerves often remain in the part of the arm that wasn't damaged. When a bionic arm is put in place, the brain can use the nerves to send commands to the arm. A computer in the arm uses the signals to make motors move the arm.

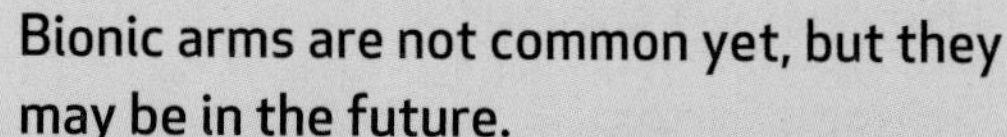

**Bionic arms are not common yet, but they may be in the future.**

**nervous system**

The **nervous system** works as a communication network to help the body respond to its environment.

Some people cannot hear because of a problem with a part of the ear called the cochlea. The cochlea's role is to pick up sound vibrations and send them along nerves to the brain. When a patient's cochlea cannot do this, doctors can implant small electrodes in the cochlea. The patient also wears a microphone. The microphone picks up sounds and sends signals to the electrodes, which send them to the nerves. The nerves send the sounds to the brain and the person hears!

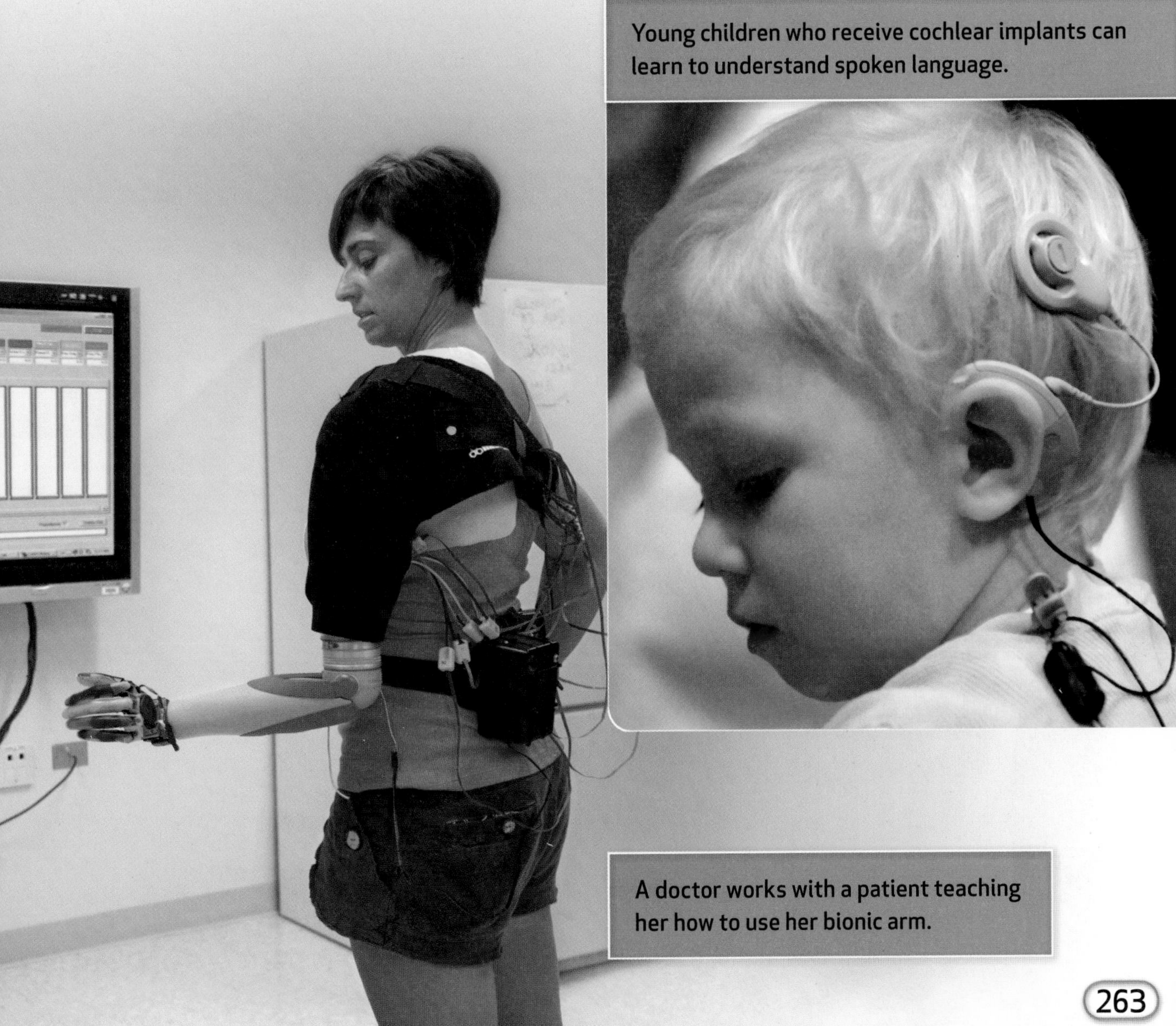

Young children who receive cochlear implants can learn to understand spoken language.

A doctor works with a patient teaching her how to use her bionic arm.

## Helping the Heart

The heart has a key role in the **circulatory system**. A patient whose heart is not working well may need a heart transplant. Often, a real heart cannot be found right away. Doctors may give patients a machine to help the heart pump blood or an artificial heart that does the job of the real heart. These machines can't take the place of a heart transplant, but they keep many patients healthy while they wait for a heart to become available. In the future, scientists hope to create an artificial heart that will work as well as a transplanted heart.

An artificial heart can do many of the same jobs as a real heart.

**circulatory system**

The **circulatory system** is a group of organs that carries blood throughout the body.

Help is also available for the **respiratory system**. In a working respiratory system, the lungs take carbon dioxide out of the blood and put oxygen into the blood. For people with lung problems, the blood circles out of a patient's body and through an artificial lung. The artificial lung adds oxygen and removes carbon dioxide from the body just like real lungs do.

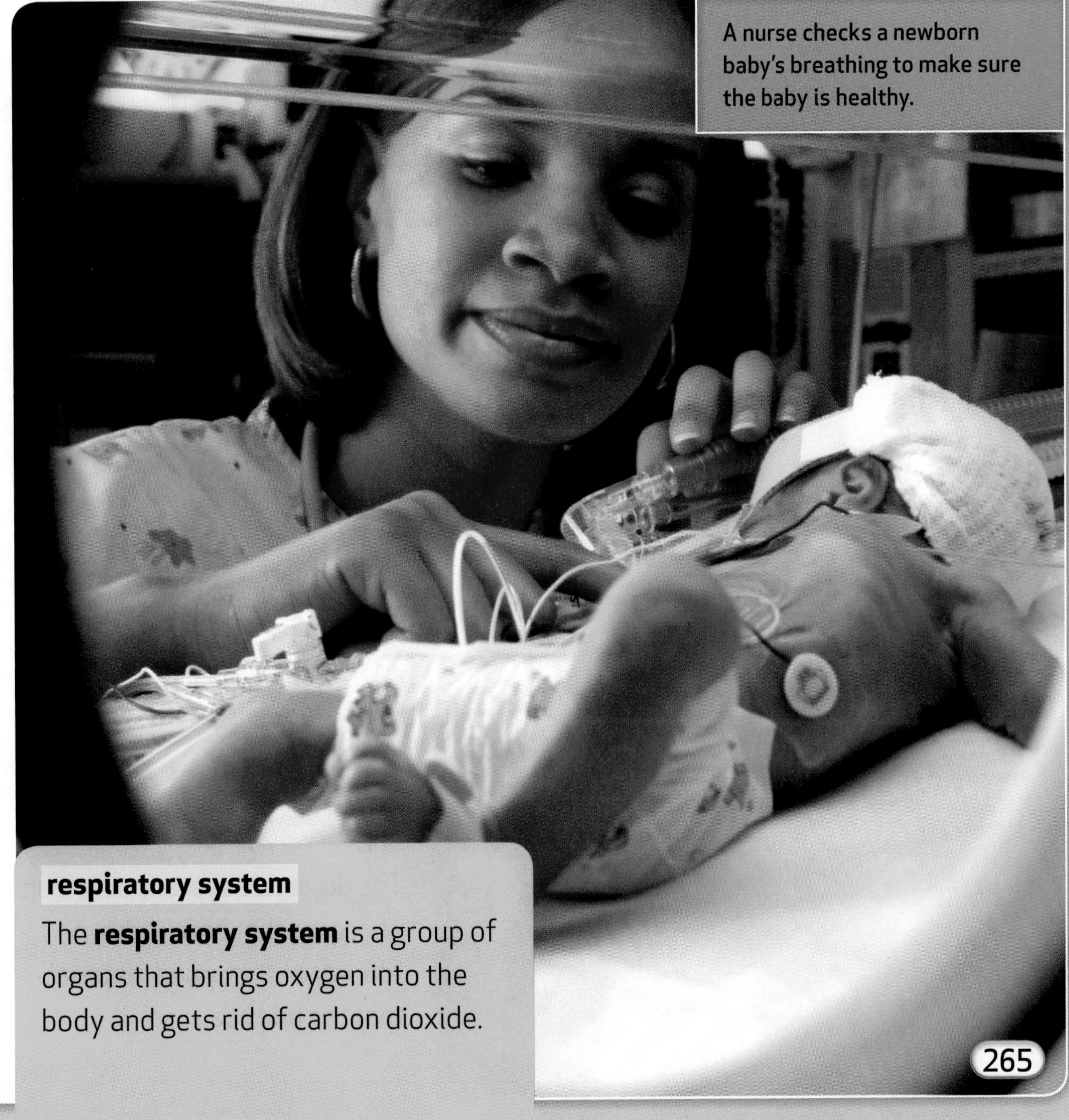

A nurse checks a newborn baby's breathing to make sure the baby is healthy.

**respiratory system**

The **respiratory system** is a group of organs that brings oxygen into the body and gets rid of carbon dioxide.

## A Look Inside

Doctors sometimes need ways to look inside a patient's **digestive system**. The doctor might use a tool with a thin lighted tube with a lens on the end. The tube is fed into the patient's body, and the doctor watches a screen to see what the lens sees. Using this tool, doctors can detect blockages and other problems in the digestive system.

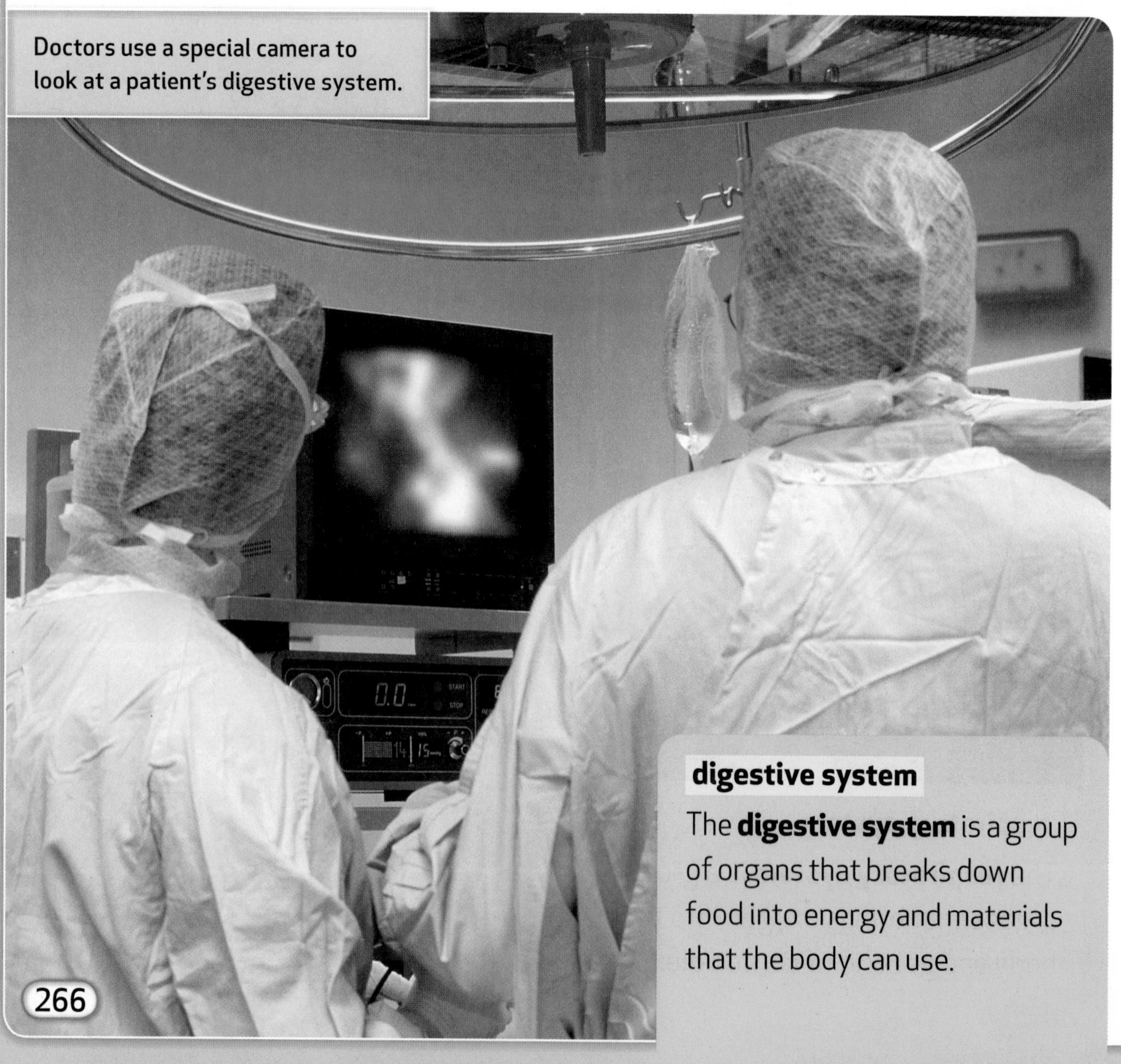

Doctors use a special camera to look at a patient's digestive system.

**digestive system**

The **digestive system** is a group of organs that breaks down food into energy and materials that the body can use.

Doctors use many kinds of machines to look more closely at the human body. Some use x-ray images and a computer to check the density of a patient's bones. MRI (magnetic resonance imaging) machines help doctors get a good look at the brain and the spinal cord. CT (computed tomography) scans take hundreds of x-ray images of a patient's body and put them together to create a three-dimensional picture. This type of scan has even been used to examine the bodies of ancient mummies!

Scientists study a 3-D image of a human skull.

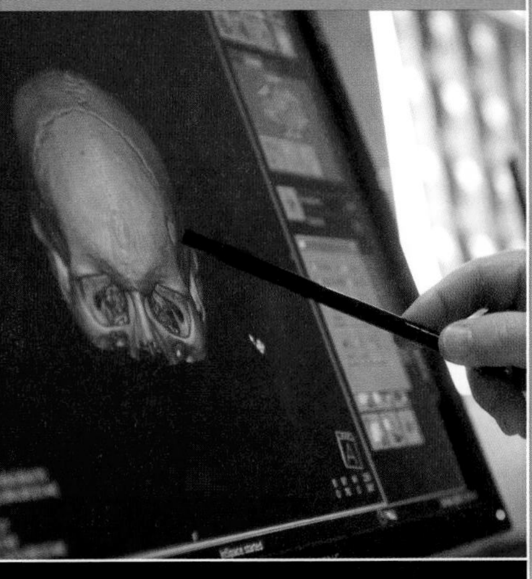

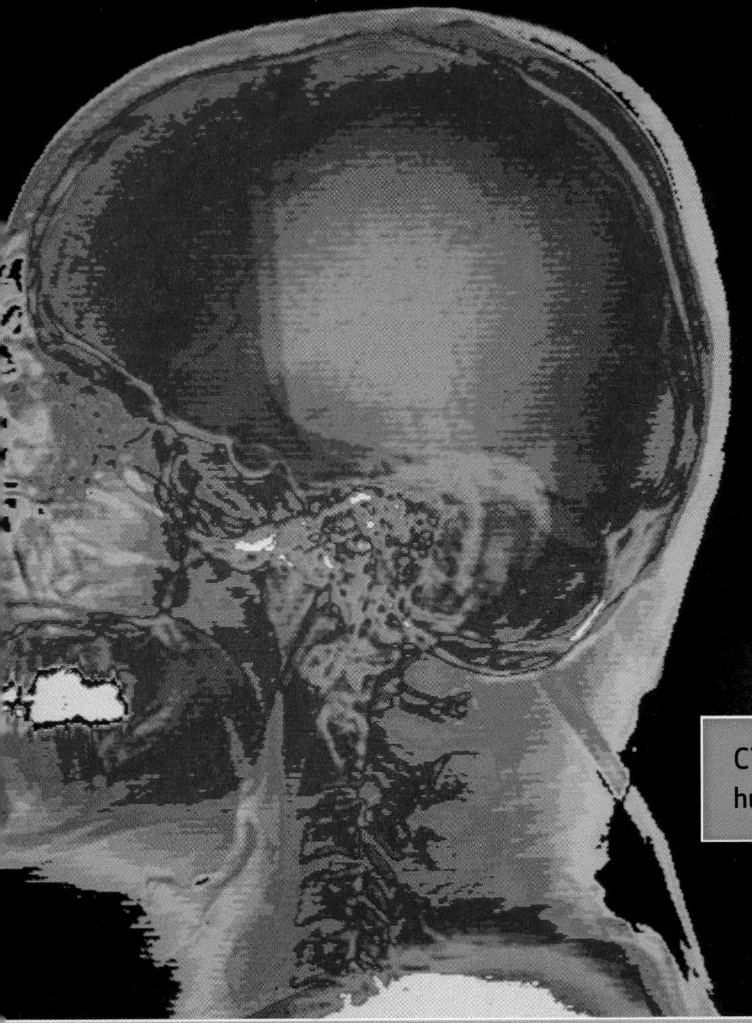

TECHTREK
myNGconnect.com

Digital Library

CT image of a human skull

NATIONAL GEOGRAPHIC

# CHAPTER 6 SHARE AND COMPARE

**Turn and Talk** How can technology be used to help keep a patient's body parts and systems working well? Form a complete answer to this question together with a partner.

**Read** Select two pages in this section. Practice reading the pages. Then read them aloud to a partner. Talk about why the pages are interesting.

my SCIENCE notebook **Write** Write a conclusion that summarizes what you learned about technology and human systems. State what you think is the Big Idea of this section. Share what you wrote with a classmate. Compare your conclusions. Did you recall how prosthetic devices can help people stay healthy?

my SCIENCE notebook **Draw** Think about a sport or activity you like to do. Then think of what you would need to consider in designing a prosthetic limb for that activity. Draw your idea of what the prosthetic limb would look like. Combine your drawing with those of your classmates to make a catalog of possibilities.

# Glossary

## A

**adaptation** (a-dap-TĀ-shun)
An adaptation is a feature that helps a living thing survive in its environment. (p. 138)

## C

**camouflage** (CAM-uh-flahj)
Camouflage is a color or shape that makes a living thing hard to see. (p. 146)

This flounder uses camouflage to ambush its food.

**carnivore** (KAR-nuh-vor)
A carnivore is an animal that eats other animals to survive. (p. 95)

**circulatory system** (SIR-kyū-luh-tor-ē SIS-tum)
The circulatory system is a group of organs that carries blood throughout the body. (p. 236)

**conifer** (CON-uh-fur)
A conifer is a seed plant that reproduces with cones. (p. 12)

**conserve** (kun-SURV)
To conserve is to use resources in a careful way. (p. 205)

One way to help conserve natural resources is to recycle cans.

# D

**digestive system** (dī-JES-tiv SIS-tum)
The digestive system is a group of organs that breaks down food into energy and materials that the body can use. (p. 240)

# E

**extinction** (eck-STINK-shun)
Extinction is the complete loss of one kind of living thing. (p. 166)

# F

**fertilization** (FUR-til-i-ZĀ-shun)
In fertilization an egg and a sperm cell join. (p. 17)

# H

**herbivore** (UR-buh-vor)
An herbivore is an animal that eats plants to survive. (p. 94)

**heredity** (hah-RED-ah-tē)
Heredity is the passing of traits from parents to their offspring. (p. 60)

Markings, such as stripes or spots, are part of an animal's heredity.

# I

**inherited** (in-HAIR-it-ed )
An inherited characteristic is passed down from parents to offspring. (p. 32)

**invasive organism** (in-VĀ-siv OR-guh-niz-uhm)
An invasive organism is a plant or animal that does not belong in a place and harms it. (p. 194)

# L

**larva** (LAR-va)
A larva is a young animal with a body form very different from the adult. (p. 55)

This larva will become an adult moth that flies.

# M

**metamorphosis** (met-ah-MOR-fah-sis)
Metamorphosis is a series of major changes in an animal's body form during its life cycle. (p. 54)

# N

**nervous system** (NER-vus SIS-tum)
The nervous system works as a communications network to help the body respond to its environment. (p. 246)

**nymph** (nimf)
A nymph is the stage in which the young animal looks like the adult. (p. 56)

## O

**omnivore** (AHM-ni-vor)
An omnivore is an animal that eats plants and other animals to survive. (p. 96)

**organ** (OR-gun)
An organ is a body part that does a specific job in an organism. (p. 234)

## P

**photosynthesis** (FŌ-tō-SIN-thuh-sis)
In photosynthesis, plants use the energy of sunlight to make food. (p. 11)

**pollination** (POL-uh-NĀ-shun)
Pollination is the movement of pollen from a stamen to a pistil or from a male cone to a female cone. (p. 16)

**pollution** (pah-LŪ-shun)
Pollution is any harmful substance that people put into the air, water, and soil. (p. 208)

**predator** (PRED-uh-tur)
A predator is an animal that hunts other animals for food. (p. 98)

**prey** (PRĀ)
Prey are animals that other animals hunt for food. (p. 98)

**pupa** (PŪ-pah)
A pupa is the stage in which the body form of a young animal changes from the larva to the adult. (p. 55)

## R

**recycle** (rē-SĪ-kuhl)
To recycle is to use something again. (p. 205)

**respiratory system**
(RES-pur-uh-tor-ē SIS-tum)
The respiratory system is a group of organs that brings oxygen into the body and gets rid of carbon dioxide. (p. 239)

## S

**seed dispersal** (SĒD di-SPUR-suhl)
In seed dispersal, the seeds of a plant are carried to a new place. (p. 20)

**system** (SIS-tum)
A system is a group of organs that works together in an organism to help it live and grow. (p. 235)

## V

**variation** (vair-ē-Ā-shun)
A variation is a different form of a feature of the same kind of living thing. (p. 149)

**One kind of seed dispersal is when animals carry the seed to a new place.**

# Index

# Credits

**Front Matter** **About the Cover** (b inset) Tim Laman/National Geographic Image Collection. (bg, t inset) David Doubilet/National Geographic Image Collection. **ii–iii** guichaoua/Alamy Images. (bl) Joel Sartore/National Geographic Image Collection. (br) Doug Allan/Photolibrary/Getty Images. (c) Luke Dollar. (tl) National Geographic Image Collection. (tr) Luke Dollar. **iv–v** Herbert Kratky/Shutterstock. **ix** (t) Kevin Dodge/Corbis. **vii** (t) Roy Toft/National Geographic Image Collection. **viii–ix** (bg) Michael Patrick O'Neill/Alamy Images. **vi–vii** (bg) D. Hurst/Alamy Images. **x–1** Christian Ziegler/National Geographic Image Collection. **2** (b inset) Alaska Stock Images/National Geographic Image Collection. (c inset) DLILLC/Corbis. (t inset) John Anderson/Alamy Images. **2–3** (bg) Dietmar Nill/Minden Pictures. **3** (b inset) Mike Kemp/Getty Images. (c inset) Alaska Stock Images/National Geographic Image Collection. (t inset) Purestock/Getty Images. **4** (bg) National Geographic Image Collection. (inset) Luke Dollar.

**Chapter 1** **5** Kristina Stasiuliene/Shutterstock. **6–7** (ChapOp) Kristina Stasiuliene/Shutterstock. **8** (b) Michael & Patricia Fogden/Minden Pictures. (c) Fred Whitehead/Animals Animals. (t) Chris Stein/The Image Bank/Getty Images. **9** (b) D. Hurst/Alamy Images. (c) Brian A Jackson/Shutterstock. **10** (inset) Tim Robberts/Stone/Getty Images. **10–11** (bg) Chris Stein/The Image Bank/Getty Images. **12–13** (bg) Kjell Brynildsen/iStockphoto. **13** (inset) Peter Arnold, Inc./Alamy Images. **14–15** (bg) John Anderson/Alamy Images. **15** (inset) MicroScan/Phototake/Alamy Images. **16–17** (bg) Michael & Patricia Fogden/Minden Pictures. **18–19** (c) PhotoDisc/Getty Images. **19** (bl) PhotoDisc/Getty Images. (br) Jaroslaw Grudzinski/Shutterstock. (tl) PhotoDisc/Getty Images. (tr) PhotoDisc/Getty Images. **20–23** (bg) Susan Gary/Flickr/Getty Images. **21** (bl) Indigo Photo Agency/Alamy Images. (br) Le Do/Shutterstock. (tl) mdd/Shutterstock. (tr) Brian A Jackson/Shutterstock. **22–23** (bg) DAJ/Getty Images. **24** (inset) PhotoStock-Israel/Alamy Images. (tc) Wildlife GmbH/Alamy Images. (tl) Wildlife GmbH/Alamy Images. (tr) Wildlife GmbH/Alamy Images. **24–25** (bg) Mircea Bezergheanu/Shutterstock. **25** (l) Terje Rakke/Image Bank/Getty Images. (r) Maryann Frazier/PSU/Photo Researchers, Inc. **26–27** (bg) Peter Chambers/Alamy Images. **27** (b) Emma Lee/Life File Photo Library Ltd/Alamy Images. (c) Christa Knijff/Alamy Images. (l) 1001nights/iStockphoto. (r) Linda Phillips/Photo Researchers, Inc. (t) Nigel Cattlin/Visuals Unlimited. **28–29** (bg) John Serrao/Photo Researchers, Inc. **29** (b) Greg Wright/Alamy Images. (l) Fred Whitehead/Animals Animals. (r) Ned Therrien/Visuals Unlimited. (t) David Sieren/Visuals Unlimited. **30** (bg) Fred Whitehead/Animals Animals. (inset) Top-Pics TBK/Alamy Images. **32–33** (bg) D. Hurst/Alamy Images. **33** (inset) ScottsMiracle-Gro. **34** (c) Top-Pics TBK/Alamy Images. (l) Nigel Cattlin/Visuals Unlimited. (r) Thomas & Pat Leeson/Photo Researchers, Inc. **34–35** (bg) Yashuhiro Matsuoka/Getty Images. **35** (inset) Jupiterimages/Getty Images. **36** (b) Ralph Lee Hopkins/National Geographic Image Collection. (bg) imagewerks/Getty Images. (t) Jim Webb/National Geographic Image Collection. **37** (bg) Jim Webb/National Geographic Image Collection. (inset) Craig Lovell/Corbis. **38–39** (bg) vito arcomano/Marka/age fotostock. **39** Jeri Gleiter/Getty Images. **40** (l) Jeff Foott/Discovery Channel Images/Getty Images. (r) Russell Wood/Visuals Unlimited. **41** (bg) Joe Austin Photography/Alamy Images. (l) Steve Matson. (r) Murray, Patti/Animals Animals. **43** (b) Mark Stouffer/Animals Animals. (t) W. Robert Moore/National Geographic Image Collection. **44** Jeri Gleiter/Getty Images.

**Chapter 2** **45, 46–47** Klaus Nigge/National Geographic Image Collection. **48** (b) Gerald J. Lenhard, Louiana State Univ, Bugwood.org/USDA Forest Service Archives. (c) Don Farrall/Photolibrary. (t) Corbis/Photolibrary. **49** (b) C.C. Lockwood/Animals Animals. (t) Robert Marien/Corbis Premium RF/Alamy Images. **50** (inset) devi/Shutterstock. **50–51** (bg) Christopher Elwell/Shutterstock. **52–53** (bg) Flip Nicklin /Minden Pictures/National Geographic Image Collection. **53** (b) Mark Conlin/Alamy Images. (l) Joe White/Shutterstock. (t) A & J Visage/Alamy Images. **54** (b) Gerald J. Lenhard, Louiana State Univ, Bugwood.org/USDA Forest Service Archives. (l) Corbis/Photolibrary. (r) Don Farrall/Photolibrary. (t) James Robinson/Animals Animals. **55** Rolf Nussbaumer/Alamy Images. **56–57** (bg) Thomas Marent/Minden Pictures. **57** (b) Robert Marien/Corbis Premium RF/Alamy Images. (l) Digital Vision/Getty Images. (t) Pfefferle M /Arco Images/age fotostock. **58** (inset) Millard H. Sharp/Photo Researchers, Inc. **59** (bg) Millard H. Sharp/Photo Researchers, Inc. **60** DLILLC/Corbis. **61** (b) blickwinkel/Alamy Images. (t) C.C. Lockwood/Animals Animals. **62** (inset) Scott Leslie/Minden Pictures. **62–63** (bg) Thomas C Kline, Jr/Alaska Stock. **64** Anup Shah/Nature Picture Library. **65** Erick Nguyen/Alamy Images. **66** (bg) Pete Oxford/Minden Pictures. (inset) Pete Oxford/Minden Pictures. **67** Pete Oxford/Minden Pictures. **68** (inset) Chris Cheadle/All Canada Photos/Getty Images. **68–69** (bg) George Doyle/Stockbyte/Getty Images. **69** (l) blickwinkel/Alamy Images. (r) Paul Doyle/Photofusion Picture Library/Alamy Images. **70–71** (bg) George D. Lepp/Corbis. **71** (l) Don Farrall/Photolibrary. (r) Gerald J. Lenhard, Louiana State Univ, Bugwood.org/USDA Forest Service Archives. **72** (b) Janet Hicks/Shutterstock. (inset) Elizabeth Howard/University of Minnesota Monarch Lab/Karen S. Oberhauser. **73** (bg) Joel Sartore/National Geographic Image Collection. (inset) Michelle Solensky/University of Minnesota Monarch Lab/Karen S. Oberhauser. **74** (inset) Biosphoto/Koenig Christian/Peter Arnold, Inc. **74–75** (bg) Marevision/age fotostock. **76–77** (bg) Hans Christoph Kappel/Nature Picture Library. **77** (bl) Allan Bergmann Jensen/Alamy Images. (br) Wildlife/Peter Arnold, Inc. (tl) Zheltyshev/Shutterstock. (tr) Wikipedia the Free Encyclopedia. **78** (inset) Martin Fowler/Alamy Images. **78–79** (bg) Bernard Castelein/Nature Picture Library. **80** (l) John Daniels/Ardea.com. (r) Brian Bevan/Ardea.com. **80–81** (bg) Robin Hill/Digital Vision/Getty Images. **81** (l) Robert Pickett/Papilio/Alamy Images. (r) OSF/Animals Animals. **82–83** (bg) Digital Vision/Getty Images. **83** (b) Chris Mattison/Alamy Images. (tl) Mark Moffett/National Geographic Image Collection. (tr) Thomas Marent/Minden Pictures. **84** Robert Pickett/Papilio/Alamy Images.

**Chapter 3** **85** Fritz Polking/FLPA/Minden Pictures/National Geographic Image Collection. **86–87** (ChapOp) Fritz Polking/FLPA/Minden Pictures/National Geographic Image Collection. **88** (b) T. J. Rich/Nature Picture Library. (t) AMA/Shutterstock. **89** (c) Dietmar Nill/Minden Pictures. (tl) Fritz Polking/Visuals Unlimited. (tr) Norbert Wu/Minden Pictures/National Geographic Image Collection. **90–91** Su Garfinkle/Axiom Photographic Agency/Getty Images. **92** (bg) Dietmar Nill/Minden Pictures. (inset) Digital Vision/Getty Images. **93** Zhu Difeng/Shutterstock. **94–95** (bg) AMA/Shutterstock. **95** (inset) T. J. Rich/Nature Picture Library. **96** (bg) Fritz Polking/Visuals Unlimited. (inset) Norbert Wu/Minden Pictures/National Geographic Image Collection. **97** iStockphoto. **99** (bg) Juniors Bildarchiv/Photolibrary. (tl) William Berry/Shutterstock. **100–101** (bg) Andrei Rybachuk/Shutterstock. **102** (l) Paul Stinsa. (r) Paul Stinsa. **103** (bg) AlaskaStock. (inset) Paul Stinsa. **104** (inset) Carl & Ann Purcell/Corbis. **104–105** (bg) Juniors Bildarchiv/Alamy Images. **106** (bg) Juniors Bildarchiv/Alamy Images. (inset) Steve Gschmeissner/Photo Researchers, Inc. **107** Maria & Bruno Petriglia/Photo Researchers, Inc. **108–109** (bg) Herbert Kratky/Shutterstock. **109** (b) Joe McDonald/Visuals Unlimited. (t) Roy Toft/National Geographic Image Collection. **110** Melissa Farlow/National Geographic Image Collection. **111** Photolibrary/Getty Images. **112** (inset) Winfield Parks/National Geographic Image Collection. **112–113** (bg) Layne Kennedy/Corbis. **114–115** (bg) Daniel Cox/Photolibrary. **115** (l) Adam Jones/Visuals Unlimited. (tr) Rui Saraiva/iStockphoto. **116–117** (bg) Alaska Stock Images/National Geographic Image Collection. **117** (l) Creatas/Jupiterimages. (r) George F. Mobley/National Geographic Image Collection. **118–119** (bg) Visuals Unlimited/Corbis. **119** (inset) Jeff Rotman/Jeff Rotman Photography. **120–121** (bg) moodboard/Corbis. **122–123** K. Yoganand/National Geographic Image Collection. **124–125, 126–127** Christian Ziegler/National Geographic Image Collection. **127** (inset) Nick Upton/Nature Picture Library. **128** Bruce Dale/National Geographic Image Collection. **129** (bg) Bruce Dale/National Geographic Image Collection. (inset) Christian Ziegler/National Geographic Image Collection. **130** Otis Imboden/National Geographic Image Collection. **131–132** Christian Ziegler/National Geographic Image Collection.

**Chapter 4** **133** Darren Matthews/Alamy Images. **134–135** (ChapOp) Darren Matthews/Alamy Images. **136** (b) Darlyne A. Murawski/National Geographic Image Collection. (t) Michael Patrick O'Neill/Alamy Images. **137** (b) Corbis. (t) Christian Ziegler/National Geographic Image Collection. **138–139** (bg) Michael Patrick O'Neill/Alamy Images. **139** (inset) Christian Heinrich/imagebroker/Alamy Images. **140** (bg) Stephen Dalton/Photo Researchers, Inc. (inset) Mattias Klum/National Geographic Image Collection. **141** (b) Stephen Dalton/Minden Pictures. (b) E. Widder/HBOI/Visuals Unlimited. (t) Dietmar Nill/Foto Natura/Minden Pictures/National Geographic Image Collection. **142** (bg) Theo Allofs/Corbis. (inset) PhotoDisc/Getty Images. **143** (b) Don Despain/www.rekindlephoto.com/Alamy Images. (bc) Anke van Wyk/Shutterstock. (t) Matthew Zhang/National Geographic Image Collection. (tc) DLILLC/Corbis. **144** (inset) H. Tuller/Shutterstock. **144–145** (bg) Stacy Gold/National Geographic Image Collection. **145** (b) Maksym Gorpenyuk/Shutterstock. (bc) Creatas/Jupiterimages. (t) Armin Floreth/imagebroker/Alamy Images. (tc) Frank Leung/iStockphoto. **146** (bg) Norbert Wu/Minden Pictures/National Geographic Image Collection. (inset) John Anderson/Alamy Images. **147** Darlyne A. Murawski/National Geographic Image Collection. **148** (inset) J. Sneesby/B. Wilkins/Getty Images. **148–149** (bg) Juniors Bildarchiv/Alamy Images. **149** (t) Christian Ziegler/National Geographic Image Collection. **150** (inset) Mitsuaki Iwago/Minden Pictures/National Geographic Image Collection. **150–151** (bg) John Foxx Images/Imagestate. **151** (inset) mlorenz/Shutterstock. **152** (inset) David M. Dennis/Animals Animals. **152–153** (bg) Wildlife GmbH/Alamy Images. **153** (inset) Ken Catania/Visuals Unlimited. **154–155** (bg) Tui De Roy/Minden Pictures. **155** (inset) Tom Brakefield/Stockbyte/Getty Images. **156** (bg) Michael S. Quinton/National Geographic Image Collection. (inset) Terry Mapstone/iStockphoto. **157** (bg) Mitsuhiko Imamori/Minden Pictures/National Geographic Image Collection. (inset) Taylor S. Kennedy/National Geographic Image Collection. **158** (inset) guichaoua/Alamy Images. **158–159** (bg) Wildlife GmbH/Alamy Images. **159** (b) Schmidbauer/blickwinkel/Alamy Images. (t) Schmidbauer/blickwinkel/Alamy Images. **160** (bg) RJN Fletcher/Alamy Images. (inset) Christian Lucas Sangoyo/Shutterstock. **161** Sau Hian Ho/Alamy Images. **162–163** (bg) Roy Toft/National Geographic Image Collection. **163** (b) tomas del amo/Shutterstock. (c) Scott Camazine/Photo Researchers, Inc. (t) Rolf Klebsattel/Shutterstock. **164** (b) Ned M. Seidler/National Geographic Image Collection. (t) Robyn Mackenzie/Shutterstock. **165** (bg) S.Borisov/Shutterstock. (inset) Aiyana Paterson-Zinkand/iStockphoto. **166–167** (bg) DLILLC/Corbis. **167** (l) Corbis. (r) Pat Canova/Index Stock Imagery/Photolibrary. **168** (inset) David Evans/National Geographic Image Collection. **168–169** (bg) Carsten Peter/National Geographic Image Collection. **169** (inset) Vanessa Lefort/National Geographic Image Collection. **170** (inset) Tim Fitzharris/Minden Pictures/National Geographic Image Collection. **170–171** (bg) Purestock/Getty Images. **171** (l) Frans Lanting/National Geographic Image Collection. (r) Mitshuhiko Imamori/Minden Pictures/National Geographic Image Collection. **172** (b) Kate Jackson. (t) Kate Jackson. **173** (bg) Thomas Northcut/Getty Images. (inset) Kate Jackson. **174** (inset) Zig Leszczynski/Animals Animals. **174–175** (bg) Ryan McVay/Getty Images. **175** (inset) Joel Sartore/Getty Images. **176** (bg) AfriPics.com/Alamy Images. (inset) Karl H. Switak/Photo Researchers, Inc. **177** (l) Michael & Patricia Fogden/Corbis. (r) Ron Kimball/Kimballstock. **178** (bg) Michael and Patricia Fogden/Minden Pictures/National Geographic Image Collection. (inset) Michael Nichols/National Geographic Image Collection. **179** Gerry Ellis/Minden Pictures/National Geographic Image Collection. **180** Joel Sartore/Getty Images.

**Chapter 5** **181, 182–183** Michael Melford/National Geographic Stock/National Geographic Image Collection. **184** (b) Ashley Cooper/Corbis. (t) Stephen St. John/National Geographic Image Collection. **185** (b) Martin Harvey/Corbis. (t) Katrina Brown/Alamy Images. **186–187** (bg) Vibrant Image Studio/Shutterstock. **187** (l inset) blickwinkel/Koenig/Alamy Images. (r inset) Rob Crandall/Alamy Images. **188** (inset) Alan D. Carey/PhotoDisc/Getty Images. **188–189** (bg) Peter Weimann/Picture Press/Photolibrary. **189** (l inset) Leonard Lee Rue III/Photo Researchers, Inc. (r inset) Derrick Hamrick/Rolf Nussbaumer Image Library/Alamy Images. **190** Stephen Krasemann/Stone/Getty Images. **191** Tim Chapman/Liaison/Getty Images News/Getty Images. **192–193** (bg) M. Timothy O'Keefe/Alamy Images. **193** (inset) John Lemker/Animals Animals. **194** (bg) Robert E. Pelham/Photoshot. (inset) Ewen Cameron/Shutterstock. **195** Stephen St. John/National Geographic Image Collection. **196** (inset) Jason Kasumovic/iStockphoto. **196–197** (bg) Tim Fitzharris/Minden Pictures/National Geographic Image Collection. **197** (b inset) Gerald & Buff Corsi/Visuals Unlimited. (t inset) photoBlueIce/iStockphoto. **198–199** (bg) Alaska Stock Images/National Geographic Image Collection. **199** (inset) Richard Du Toit/Minden Pictures. **200–201** (bg) George D. Lepp/Corbis. **201** (inset) Lori Oberhofer, National Park Service/University of Florida. **202** (inset) Teakle Collection/Northwest Room, Spokane Public Library. **202–203** (bg) Brent Jolly/Alamy Images. **203** (inset) Mark Turner/age fotostock. **204** John Humble/Getty Images. **205** Ashley Cooper/Corbis. **206** (inset) Grant Heilman/Grant Heilman Photography. **206–207** (bg) Daniel Carapellotti/Getty Images. **207** (inset) Kevin Steele/Aurora Photos/Corbis. **208** (l inset) Martin Harvey/Corbis. (r inset) Photoshot Holdings Ltd/Alamy Images. **208–209** (bg) Natalie Fobes/Corbis. **209** (l inset) Olivier Blondeau/iStockphoto. (r inset) Katrina Brown/Alamy Images. **210** (bg) John W. Bova/Photo Researchers, Inc. (inset) Hal Horwitz/Corbis. **211** Chris & Tilde Stuart/Minden Pictures. **212** (bg) Dennis Frates/Alamy Images. (inset) Maximilian Weinzierl/Alamy Images. **213** Mark Conlin/Alamy Images. **214–215** (bg) Wally Eberhart/Visuals Unlimited. **215** (inset) Kevin Steele/Aurora Photos/Corbis. **216** (bg) PhotoDisc/Getty Images. (inset) Dennis Farrell/AP Images. **217** Rolf Bettner/Severn Cullis-Suzuki. **218–219** (bg) Tim Pannell/Corbis. **219** (inset) Franklin Viola/Animals Animals. **220** Wendy Stone/Corbis. **221** Wendy Stone/Corbis. **222** (bg) Jerry & Marcy Monkman/Animals Animals. (inset) Phyllis Greenberg/Animals Animals. **223** Lon Lauber/Alaska Stock **224** David Northcott/Danita Delimont Stock Photography. **224–225** Jason Edwards/National Geographic Image Collection. **226** Alex Webb/National Geographic Image Collection. **227** Andreanna Seymore/Getty Images. **228** Phyllis Greenberg/Animals Animals.

**Chapter 6** **229** Carson Ganci/Design Pics/Corbis. **230–231** Carson Ganci/Design Pics/Corbis. **232** (c) Werran/Ochsner/Getty Images. **234–235** (bg) Werran/Ochsner/Getty Images. **235** (inset) Stockbyte/Getty Images. **236–237** (bg) Alistair Berg/Getty Images. **238–239** (bg) Peter Cade/Getty Images. **240** Mike Kemp/Getty Images. **242–243** (bg) Loungepark/Getty Images. **244–245** (bg) Mike Powell/Getty Images. **245** (inset) Anthony-Masterson/Getty Images. **247** Kuzma/Shutterstock. **248–249** (bg) SW Productions/Getty Images. **249** (inset) Bananastock/age fotostock. **250** (bg) Elena Elisseeva/Shutterstock. (inset) Dennis Kunkel Microscopy, Inc./Visuals Unlimited. **251** Kevin Dodge/Corbis. **252** (b) Image by Barry Forbes/Getty Images. (t) Westend61 GmbH/Alamy Images. **253** Bernd Thissen/epa/Corbis. **254** (inset) Purestock/Getty Images. **254–255** (bg) Marcus Armani/National Geographic Image Collection. **256–257** (bg) Jill Sabella/Getty Images. **257** (inset) Werran/Ochsner/Getty Images. **258** (b) M. Dores. **258–259** (tc) CNRI/SPL/Getty Images. **259** (b) Sebastian Kaulitzki/Shutterstock. (tr) CNRI/SPL/Getty Images. **260** Chase Jarvis/Getty Images. **261** (bg) Simon Bruty/Sports Illustrated/Getty Images. (inset) Simon Bruty/Sports Illustrated/Getty Images. **262** (inset) Mark Thiessen/National Geographic Image Collection. **262–263** (bg) Mark Thiessen/National Geographic Image Collection. **263** (inset) Gene J. Puskar/AP Images. **264** (bg) Phanie/Photo Researchers, Inc. (inset) Henning Dalhoff /Bonnier Publications/Photo Researchers, Inc. **265** David Joel/Getty Images. **266** Keene/BSIP/age fotostock. **267** (bg) Mira/Alamy Images. (inset) Luther Caverly/Brownstock Inc./Alamy Images. **268** Simon Bruty/Sports Illustrated/Getty Images.

**End Matter** **EM1** (bg) Tim Pannell/Corbis. (inset) John Anderson/Alamy Images. **EM2** (l) C.C. Lockwood/Animals Animals. (r) Don Farrall/Photolibrary. **EM3** Susan Gary/Flickr/Getty Images. **EM5** Rolf Klebsattel/Shutterstock. **EM9** DAJ/Getty Images. **Back Cover** (bg) David Doubilet/National Geographic Image Collection. (bl) Joel Sartore/National Geographic Image Collection. (br) Doug Allan/Photolibrary/Getty Images. (c, tr) Luke Dollar. (tl) National Geographic Image Collection.